SIX MONTHS IN THE NEW GOLD-DIGGINGS

John Lyle Campbell, 1818-1886, the author of this book, was a reporter for the *Chicago Tribune*. His book was first published in 1864 (written in 1863) by the Tribune Book and Job Office, 51 Clark Street, Chicago, Illinois. It was also printed later that same year in New York, and was printed again in Chicago in 1866. There were several variations in title pages and number of appended advertisements, making this rare book either the despair or delight of book collectors. Campbell also wrote a number of shorter items and is a possible author of the anonymous *Banditti of the Rocky Mountains*, according to Jerome Peltier, p. 17 of the reprint of this title as published in 1964 by Ross & Haines of Minneapolis. The original *Banditti* was printed in 1865 by Wilson and Company of New York, and by "Box 3197" of Chicago.

The original title of the Campbell book on placer gold mining in Idaho Territory in 1863 was *Idaho: or Six Weeks in the New Gold-Diggings.* As an Idaho title it is reasonable to expect to find some copies in the PNW, however the 1950 edition of Charles W. Smith's *Pacific Northwest Americana* locates only three copies in the Pacific Northwest. Portland Public Library holds the two Chicago editions of 1864 and 1866. The University of Oregon at Eugene holds the 1864 New York edition.

SIX MONTHS IN THE NEW GOLD-DIGGINGS

PLACER GOLD MINING IN IDAHO TERRITORY IN 1863.

JOHN L. CAMPBELL

YE GALLEON PRESS
FAIRFIELD, WASHINGTON
1979

Library of Congress Cataloging in Publication Data

Campbell, John Lyle
Six months in the new gold-diggings.

Reprint of the 1864 ed. published by J. R. Walsh, Chicago, under the title: Idaho, six months in the new gold-diggings.

1. Idaho—Description and travel. 2. Gold mines and mining—Idaho. 3. The West—Description and travel—1860-1880. I. Title.

F746.C23   1979   917'96'04'2   79-23370
ISBN 0-87770-225-X

Idaho and its Gold Fields

Six Months in the New Gold-Diggings

IDAHO IS an Indian word, signifying *"The Gem of the Mountains,"* an appropriate name for the attractive region which, by the organic act of Congress, dated March 3rd, 1863, wears this appellation. Nestling in the interior of the continent, traversed by the rocky chain that forms its backbone, the beautiful region is rich in its agricultural resources, favored in its climate, and now invites the eager gold-hunter by its shining ores, which until now have been locked fast as among the last surrendered secrets of the country. Idaho is indeed the hoarded gem that civilization is speedily to win and wear among its fairest jewels. A reference to the map that fronts this little volume gives the location of Idaho, as relates to the other Territories of the far West. It was carved out of large portions of three other Territories, namely, Washington, Dakota and Nebraska, thus making the chain of the Rocky mountains divide it a little west of its centre line. Idaho includes within its limits one of the richest and most interesting of all the gold fields that has yet invited the explorer, either on a golden or a scientific quest.

Copying from the organic act, the boundaries of Idaho are as follows:

"Beginning at a point in the middle channel of the Snake river where the northern boundary of Oregon intersects the same; then following down said channel of Snake river to a point opposite the mouth of the Kooskooskia or Clearwater river; thence due north to the forty-ninth parallel of latitude; thence east, along said parallel, to the twenty-seventh degree of longitude west of Washington; thence south, along said degree of longitude, to the northern boundary of Colorado Territory; thence west along said boundary, to the thirty-third degree of longitude west of Washington; thence north, along said degree, to the forty-second parallel of latitude; thence west, along said parallel, to the eastern boundary of the State of Oregon; thence north, along said boundary, to the place of beginning."

More briefly stated, the Territory is bounded north by British America, east by Dakota and Nebraska, south by Colorado and Utah, and west by Utah, Oregon and Washington.

The area thus enclosed contains three hundred and twenty-six thousand square miles, which the reader will better appreciate from the comparative statement that it is seven times as large as the State of New York, and five times the size of the six New England States.

The gold regions of Idaho owe their first discovery to the rebound of the wave of emigration that first rolled from the older States toward the Pacific coast, sheer across the continent. This vast adventurous army of restless and tireless gold-seekers first diffused themselves through the rich tracts of California and the Pacific coast, and has been working its way back thence toward the heart of the wilderness, until what was once laid down in the maps as an unknown waste save where sparsely investigated by various explorers, is becoming dotted with infant settlements by the magic wand of the enchanter Gold; while the region, for years abandoned as a wild for the Indian and the buffalo, is seeing the first frame work of civilized society laid across its whole extent, whereon will be built thriving and prosperous States ere the generation that first knew "the California fever" shall have passed away.

The Territory of Idaho is as yet unsurveyed, and only imperfectly subjected to scientific exploration; but when did the emigrant and the gold-seeker wait for line and level to establish, and the patient science of the savan to enlighten, the path along which he reads the lure of a new home and wealth in the wilderness? To open new regions himself, to press with his own feet virgin soil, and expose with his own hands new fountains of wealth, is an attraction more strong than the most careful notes of more scientific predecessors could assure him.

The purpose of the present little volume is no other than to furnish the gold-seeker who is intent on Idaho, with a few notes of residence and research in that region, with a carefully prepared itinerary of the route thither, and with such hints as to preparation and outfit as may serve the reader either in answer to merely curious inquiry, or the more earnest quest of intended journey thither.

The writer left Chicago, Illinois, on the 20th day of April, 1863, in a company which, on arriving at Fort Kearney, had increased to the number of twenty-four wagons and some seventy persons. Thus far we found the road good, and the weather most beautiful; the sky had scarcely been darkened with a cloud except on one occasion when we were drenched to the skin by a severe thunder storm.

The grass shooting forth from the effects of the rain, though scarce at the commencement of our journey, was now abundant, and our stock in a thriving condition. On the morning of the 11th of May, we broke camp at Fort Kearney, and bid adieu to civilization. Westward our course lay along the north side of the Platte, at a pace of from seventeen to twenty miles per day. Here we crossed the threshold of the great American Desert, beyond which is a dreary monotony of sandy, sterile plains extending to the very base of the Rocky mountains, broken only by an occasional bluff, or a low, irregular chain of hills.

On the 1st of June we reached Fort Laramie, which is a military post of several years standing, situated upon the south side of the Platte, at the mouth of Laramie river, and is now garrisoned with Ohio troops. The fort is composed of logs and adobe buildings.

Resuming our march we camped at the base of the Black Hills the same night. There the road leaves the river, and crosses the hills, which are a mass of cragged and broken knobs, with many steep ascents. Upon these hills we camped two different nights without water either for cooking purposes or for our stock, but as we had found plenty during the daytime, of course we did not suffer much.

From Fort Laramie to Bannock City, the country has a more wild and picturesque appearance. The famous South Pass is distant from Fort Laramie 320 miles, from St. Louis 1,580, and from the mouth of the Columbia, about 1,400. It is, therefore, nearly mid-way between the Mississippi and the Pacific. The altitude of this wonderful place is 7,490 feet above the level of the sea, and it is from twenty to thirty miles wide.

The ascent is so imperceptible that it is not an easy task to ascertain the dividing line. A stony ridge crossing the road on the table-land is designated as the spot, and its position has been fixed at N. Lat. 42° 20', and W. Long. 113°. At the north lies that noble and picturesque chain of *Les Montagnes*

Rockeuses, more commonly called Wind River mountains. On the south is situated Table mountains, an insignificant chain of low hills.

It is not a Pass, as many think. It may have some resemblance to the pass of the Alleghany, or the Barrancas of Mexico. It is not what it seems to the general reader, —a pass dividing lofty mountains, whose frowning peaks and beetling walls rise in forbidding grandeur, looking down upon the traveler, as he winds his way through a giant gate-way to the Western Continent. And yet the word *Pass* has its meaning. In America, where Nature has spread out her largest and most gorgeous map, where every feature of natural scenery—mountain and prairie, lake and river, forest and dell—dwarf their congeners in the old world, this steppe of the continent,—level-topped bluff,—upon whose surface there is room enough for the *armies* of the world to march over, is a most glorious and grand avenue. This is a suggestive spot; here, memory is irresistibly carried back to earlier scenes, before plunging into the mysterious lands of the far West. The wonderful net-work of railway, with which steam has bridged the banks of the Hudson; the soft and sunny scenery of Ohio, and the sweeping course of the mighty Mississippi; the terrible grandeur of Niagara, and the wonderful chain of placid lakes, sweeping its way from Ontario to Superior; the verdant pasture lands of the North, the rich plantations of the tropical South, and the out-stretching corn-fields of the West; lastly, the luxuriant meadow lands, and the gloomy desert waste of alkali and sage, of deep defile and crumbling bluff, like the ruins of some ancient world;—all pass before the mind era they are thrust into oblivion by the excitement of the picturesque grandeur of other regions.

At the South Pass we encamped the 22nd of June, when we took Lander's Cut-off, which is the old Oregon road, bearing a north-westerly direction.

The next thing worthy of mention, is the Bear River chain of mountains, which our route crosses by a road opened by Government, at an expense of several millions of dollars,—an achievement of the lamented Lander, in whose charge the work was completed. The road is an excellent one, and is from sixty to seventy miles in length, across the two ranges. By the consummate skill of the military engineer, easy grades have been secured, the road daringly crossing at points where the summer traveler sees snow nestled beneath him in the sunless hollows, while above him tower the snow-capped peaks that hold their winter treasures until long after the summer solstice.

From Fort Kearney to the South Pass, the greatest scarcity of the party was in the matter of fuel, though it was rare that a supply of buffalo chips could not be collected with a little pains, and serve a tolerable purpose for cooking. Grass and water we found plenty, in this portion of the route.

When we reached Lander Cut-off, we found our full lack liberally supplied, while the abundant and nourishing mountain grass, and the frequent gushing springs, kept our stock in the best condition. We lost only two head of our cattle, and these from drinking from an alkali spring. Others similarly exposed were cured by administering pieces of fat bacon, the grease of which forms the antidote by combination with the alkali, the product being soap, and the effect is to physic and relieve the animal.

We are now beyond the Bear River mountains, and have entered long since the Territory of Idaho, whose boundary we passed thirty miles east of Fort Laramie. The route is throughout the most interesting imaginable. The wonders of nature on every hand attract and delight the lover of her works, and often elicit from the dullest observer an exclamation of wonder and delight. Rough, precipitous peaks, snow-capped and glistening; narrow gorges, steep ascents, or frowning, inaccessible acclivities;—among such our way was taken, the hand of man having so tamed the wilderness that our train of wagons found scarcely more difficulty than among mountain roads in the older States.

In one of these mountain valleys, on the eastern base of the west chain, we saw a beautiful and valuable salt spring,—a spot of several acres, oozing with brine and forming a small creek of salt water, pronounced by one of our party equal to the springs of Onondaga. The banks of these springs, the rocks and exposed surfaces along the little creek, glistened with encrustations of pure virgin salt, of excellent quality, whereof bushels might readily be gathered. The future product of this spring may well vie with the richest placer of the region beyond.

Here, too, we saw what to most of us was our "first grizzly," a huge creature of this class coming at a slow canter round a bluff, right toward us. He was an immense fellow, and of course no smaller in our eyes, than his actual size. The excitement was intense. Guns and rifles were got out, and in

John L. Campbell

their eagerness some ran without their weapons, to be the first to get a shot at him. He waited for no near range, was not in the mood for a quarrel with strangers, but speedily betook himself to the fastnesses of an adjoining marsh, and we saw him no more.

Six miles further on, after the episode of Sir Grizzly, our party encamped and passed the 4th of July, in a beautiful little valley, which we named Independence Valley, with appropriate ceremonies. The Blackfoot creek, a tributary of the Bear river, which empties into Salt Lake, flows through this valley. It abounds in delicious fish, which at the time of our visit were a prey to the piscatorially inclined of our party, who with swiftly plying lines, landed nearly a barrel of plump trout and "mountain suckers" on the bank, in an incredibly short space of time. We feasted on fish, to the cooking of which we gave every variety of treatment, consistent with our limited cuisine. A number of pleasant farms are already located in this valley.

And here our route intersects with the Fort Bridger and Bannock road, the itinerary of which is given elsewhere. As we passed that way on our return, it will be appropriate to make a brief detour in this narrative, to speak of some of its features.

At South Pass, the route to Salt Lake diverges to the southward, passing Fort Bridger, a military post in the north-east corner of Utah Territory, garrisoned by California troops. Our own route from the South Pass is the Oregon road, opened by Col. Lander, as the shorter military cut-off. From Fort Bridger a route strikes across, rejoining this Oregon road at Independence Valley, above named, and in very wet seasons this roundabout, by way of Fort Bridger, is to be taken by emigrants, to escape flooded streams and other obstacles. We refer to the itinerary for distances and features of these routes. With this cross-route from Fort Bridger intersects, at Soda Springs, a route striking south-westerly the Salt Lake route. Soda Springs is thirty miles south of the Oregon road, at Independence Valley. And here the two waves of emigration and gold-seeking from east and west, meet and pour into Idaho in one channel—this, from the Atlantic slope, by way of the Platte Valley; that from California through Salt Lake.

Soda Springs is situated on Bear river, a place from 300 to 500 inhabitants. It takes its name from the alkaline springs which abound in that vicinity. Here is concentrating the Morrisite faction of the Mormons, a branch broken off from that great social Upas, by a refusal to follow the peculiar marriage tenets enforced in the realm of Brigham. The Morrisites number a little over one thousand, and they are rapidly taking up farms in the beautiful valley. They are mostly foreigners, and are led by a Scotchman named Dow. They believe in Joe Smith, but reject polygamy. They went to Soda Springs under escort of United States troops, to escape deadly persecution at Salt Lake.

Soda springs gush forth from the earth in many places, steaming with heat, depositing sulphuret of lime, until mounds are formed many feet above the level of the earth. Some of the mounds were thirty feet high, appearing like huge rocks, and one seemed to be hollow, from the sound under my feet as I walked over it. These rocks are formed in this wise; the water breaks through the surface of the earth, then gradually depositing its sulphuret of lime, when at some point of time the passages become closed, and the water is compelled to find a new outlet. We found many new springs that had recently forced a passage through the surface, as also many rocky mounds from which no water was gushing forth, the passages having been entirely closed.

Within a few miles of these springs are vast deposits of native sulphur inexhaustible in richness. The past volcanic character of this region is attested by the existence of a huge external crater which we visited, a few miles from Soda Springs, which will well bear the investigation of the scientific explorer, and deserves a larger space in description than we can give to it here.

The Bear river is a stream 400 miles in length, which is so thoroughly an institution of Utah that it actually begins and ends in that Territory, though it throws its middle portion in a wide loop into Idaho. At Soda Springs it is about fifty yards wide, generally shallow, with a rocky bottom, and its banks vary from bold bluffs to smooth levels. Its waters, fed by mountain springs, are pure, and abound in fish. About 170 miles below Soda Springs, the Bear river flows into the great salt inland sea. Fort Conner is a military post at Soda Springs, established by Capt. Black, with a detachment from Gen. Conner's force at Utah, in the summer of 1863, bringing the distressed Morrisites thither, laying out their town, settling them in their homes, and building permanent and substantial barracks for the force the Government will retain here for some time to come.

Idaho and its Gold Fields

Our course now lay northward, along the west base of the mountains. We reached Snake river, crossing it on Harry Rickard's ferry, which is the only reliable one on this portion of the river. The current of this river is swift, rushing over immense boulders, which renders fording impossible. Mr. Rickard's ferry, situated at this point, is managed by himself, and is well worthy of the patronage of the traveling public. On the following page is a cut, illustrative of the ferry and its surroundings.

The Snake river, like all other streams in this country, abounds in speckled trout, and our party caught a supply, the largest one weighing four pounds. We next recross the Rocky chain, in order to reach the head-waters of the Missouri. The crossing at this point is much like that of the South Pass, but not so wide, with only one or two short, steep ascents. Upon the summit at this point, it was our fortune, with several others, to camp, the night of the 22nd of October, 1863, while on our return to the States. The first snow of the season had whitened the summit and the surrounding hills, and the cold was so intense that one of our party froze his feet, and another froze his ears. Spreading our blankets upon the snow, we turned in for the night, with naught, save the starry heavens, to shield us from the wintry storm.

The following morning we descended toward the Snake river, some eight and a half miles, where, in a powerful contrast with our night's experience, we found a party of men busily engaged in cutting and making hay.

From this point we reached, in four days, Bannock City, the distance being seventy miles, and agricultural valleys of Eastern Idaho. As the fruits of such residence and adventure, the facts in the following pages are secured, combining there with information derived from other sources, not to omit mention of the report of Capt. Fisk, U.S.A., published early in the present year. As above stated, the first discovery of gold in the Rocky mountain region, was due to the refluence of the tide of hardy explorers penetrating farther and farther inland from the Pacific slope, returning to follow their ardent quest in regions their feet had eagerly left behind in their first emigration.

And meeting this returning wave from the Pacific, a new rush of emigration from the Atlantic slope is pouring to join it in the very heart of the continent. The mountain region, the last treasure-house of the continent, is unlocked. Not a week passes but new mines and new diggings in Idaho, and

HARRY RICKARD'S FERRY, ON SNAKE RIVER

her sister Territories, reward enterprise and enrich discoveries in these golden tracks, equally rich and many times more extensive than those of California.

It is not the purpose of this volume simply to represent the treasures of new gold-fields as the sole claim of Idaho to invite emigration, but these as the principal attraction of a freshly-opened region, where communities will thrive, fostered by all that Nature affords to nourish and bless civilized life, and where homes will not only be founded on the happy but precarious fortunes of the gold-hunter, but on more secure and wider foundations of successful labor in every department of life.

We here borrow from Capt. Fisk, whose exploration was made under the auspices of the Federal Government, and the report of whose research was made to Government.

John L. Campbell

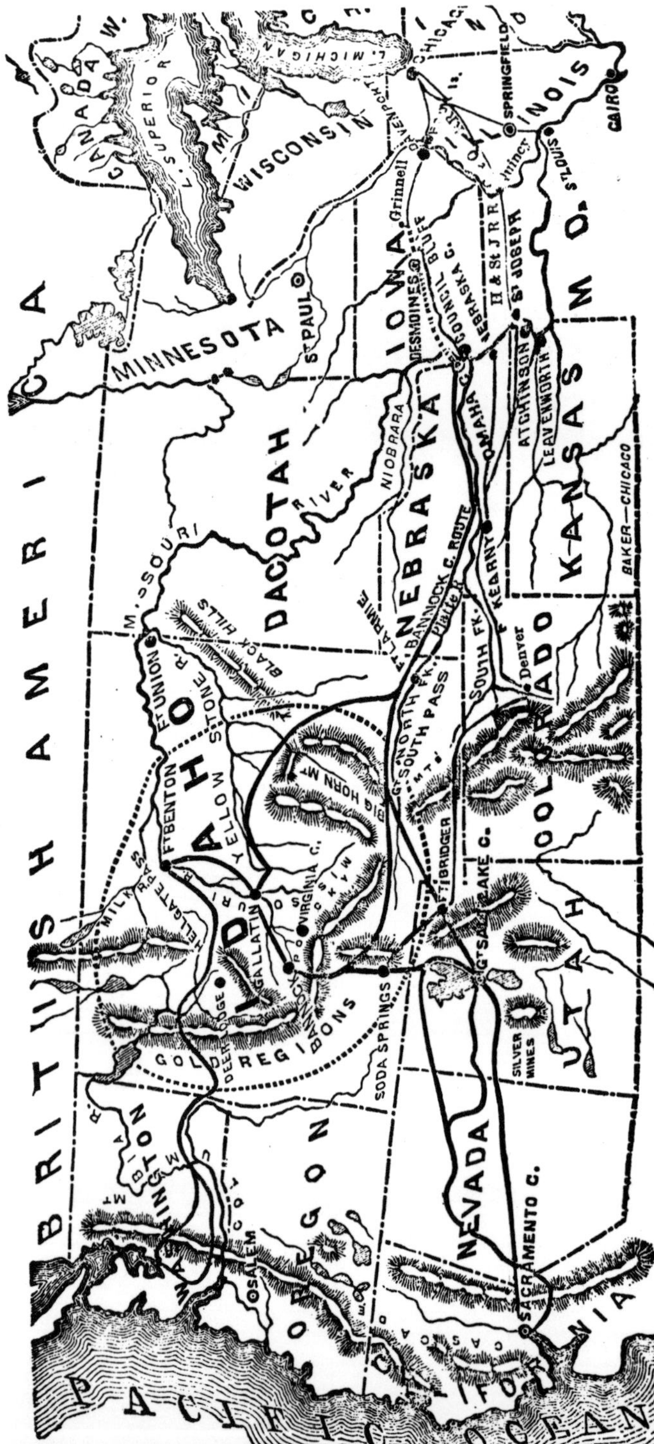

Idaho and its Gold Fields

CLIMATE AND SOIL.

The Territory of Idaho extends across so many degrees of latitude, that on this account alone it has a great variety of climate. But added to this is the difference of altitude of its different sections, from its lowest plains and valleys to the high mountain region, which contributes to the same result. The metereological statistics, as well as the experiences necessary to give a determinate and reliable statement of its climate, its heat and cold, amount of rain and snow, length of its different seasons, and a comparison of them with those of other portions of the country, are of course wanting, and will be among the first duties of the General Government to supply.

The southern portion of the Territory is mild, and from the testimony of explorers and settlers, as well as from my own experience and observation, the extreme northern portion is favored by a climate healthful to a high degree, and quite as mild as that of many of the Northern and Western States. This is particularly the case west of the mountains, in accordance with the well-known fact that the isothermal line, or the line of heat, is farther north as you go westward from the Eastern States, toward the Pacific.

At Fort Benton, on the Missouri river, in about 48° of north latitude, a trading post of the American Fur Company, their horses and cattle, of which they have large numbers, are never housed or fed in winter, but get their own living without difficulty.

The comparison which would illustrate the relation of the climate on the western border of the plains of the Missouri to those of the East, or of known points in the Mississippi valley, are first necessary, and they may be, to a certain extent, made directly with the points for the same months.

The elevation of Fort Benton above the sea, is 2,662 feet. The most conspicuous feature of the temperature at this post, is its near agreement with that of posts on the eastern border of the plains,—even in lower latitudes, as Forts Snelling and Leavenworth,—from the commencement of the record of Gov. Stevens to the close of October. For the portion of September observed, and for the whole of October, it was warmer than Fort Snelling, and but little cooler than Fort Leavenworth. The extremes of 80° for the highest, 10° for the lowest degree, are nearly the same as at Fort Snelling, where the lowest degree in October is 8°; the lowest at Fort Leavenworth being 22° and the lowest at Fort Laramie 20°.

For November, the range of temperature was lower, and on six days at or below zero. The monthly mean was ten degrees less than at Fort Snelling, and 24° less than at Fort Leavenworth. It is also 14° less than that of Cantonment Stevens on the west of the mountains. The fall of the temperature, as winter approaches, appears to be much more abrupt east of the mountains, in this latitude, than at the West, or in the vicinity of the Great Lakes. The record for December is quite anomalous, and gives an extra-ordinarily high temperature. At all other posts of the North-West—Forts Snelling, Kearney, Laramie, and east of the mountains, and at Cantonment Stevens and Olympia, on the west—the temperatures of December are several degrees below those of November, while, as recorded at Fort Benton, December is 13° warmer than November.

The record is very nearly the same as that at Fort Laramie, for the same month, though colder than this post in November, by 22° in the mean temperature. In the Deer Lodge prairie, which is along the valley of the Deer Lodge river, just west of the mountains, are as fine farming lands as can be found anywhere. The cattle here run at large in winter, and are the fattest and finest that the writer ever saw at grazing;—so fat, indeed, that I was told they always select the poorest for beef.

There is quite a settlement in this valley, and stock-raising is becoming a lucrative business, as the mining population in the vicinity is rapidly increasing, and affords a good market. At about the latitude of 46° 30', the Deer Lodge river and the Blackfoot form a junction, and are then called the Hell Gate, which unites with the Bitter Root or St. Mary's river, in latitude 47°, and assumes the name of the latter.

Along the valleys of both the Hell Gate and Bitter Root, there is a great abundance of excellent timber—pine, hemlock, tamarack or larch predominating. Beautiful prairie openings occur at frequent intervals, with good soil inviting the hand of the husbandman. At the settlement called Hell Gate, situated at the junction of the river by that name and the Bitter Root, are several farms, which yield all the cereals and vegetables in great abundance, bringing good prices, such as would astonish farmers in *the States* as parties are constantly passing through that region on their way to the mines, and glad to purchase supplies.

Idaho and its Gold Fields

At about 47^0 30', on the river by the same name, is the Coeur d'Alene Mission, a Jesuit institution, founded twenty years ago, for the purpose of converting the Indian tribes in that vicinity to Christianity. This mission is situated on the edge of an extensive and beautiful prairie, and there is a farm of one hundred and sixty acres under cultivation in connection with it. The crops raised here would astonish Eastern farmers. As high as sixty bushels of wheat to the acre are raised in favorable seasons, as I was told by the managing agent of the institution, who has been such since its foundation. There are several other missions of a similar character further to the north, where the soil produces equally well as at this, and where they raise as fine horses and cattle as can be found upon any farm in the country.

To give a more full idea of this section of country, we quote the following from the report of Governor Stevens, who had charge of an expedition that wintered in that locality. He says:

"I estimate that in the valleys on the western slopes of the Rocky mountains, and extending no farther than the Bitter Root range of mountains, there may be some 6,000 square miles of arable land, upon grassed lands with good soils, and already prepared for occupation and settlement, and that in addition to this amount, there are valleys having good soils, and favorable for settlement, which will be cleared in the removal of lumber from them. The faint attempts made by the Indians at cultivating the soil have been attended with good success, and fair returns might be expected of all such crops as are adapted to the Northern States of our country. The pasturage grounds are unsurpassed. The extensive bands of horses owned by the Flat-head Indians occupying St. Mary's village, on Bitter Root river, thrive well winter and summer. One hundred horses belonging to the exploration are wintered in this valley, and up to the 9th of March the grass was fine, but little snow had fallen, and the weather was mild. The oxen and cows owned here by the half-breeds and Indians obtain good feed, and are in good condition. Probably 4,000 square miles of tillable land is to be found immediately on the eastern slope, and the bottoms of the different streams, retaining their fertility for some distance after leaving the mountains, will considerably increase this amount. To bring out more clearly the character of the mountain region, I will, at the risk of some repetition, quote from Lieutenant Mullan's report of his exploration to Fort Hall:

"Thus we found ourselves at the main camp, after an absence of forty-five days during which time we had crossed the mountains four times turning the eastern portion of the Bitter Root which has a line of seven hundred miles, experiencing a complete change of climate, and crossing two sections of country, different in soil, formation, natural feature, capabilities, and general character; crossing, therefore, in all their ramifications, the head-waters of the two great rivers, Missouri and Columbia. We had now a fine opportunity to compare the climate and character of the Bitter Root valley, with that of the Hell Gate and others in its vicinity. In the latter, snow from four to six inches deep was to be found, while in the former the ground was perfectly free from snow. It seemed as if we had entered an entirely different region and different climate; the Bitter Root valley thus proving that it well merits the name of the valley of perennial spring. The fact of the exceedingly mild winters in this valley has been noticed and remarked by all who have ever been in it during the winter season; this affording an excellent rendezvous and recruiting station for the Indians in its vicinity and those sojourning in it, as well as others who may be overtaken by the cold or snow of the mountains. It is the home of the Flat-head Indians, who, through the instrumentality and exertions of the Jesuit priests, have built up a village—not of logs but of houses—where they repair every winter; and with this valley covered with an abundance of rich and nutritious grass, affording grazing to their large bands of horses, they live as comfortably and as happily as probably any tribe of Indians either east or west of the Rocky Mountains.

" 'The numerous mountain rivulets, tributary to the Bitter Root river, that run through the valley, afford excellent and abundant mill-seats; and the land bordering these is fertile and productive, and has been proved beyond a cavil or doubt to be well suited to every branch of agriculture. I have seen oats grown by Mr. Jonah Owen that are as heavy and as excellent as any I have ever seen in the States; and the same gentleman has informed me that he has grown most excellent wheat, and that from his experience while in the mountains he hesitates not in saying, that agriculture might be carried on here in all its numerous branches, and to the exceeding great interest, and gain of those engaged in it. The valley and mountain slopes are well timbered with an excellent growth of pine, which is equal in every respect to the well known and noted pine of Oregon. The advantages, therefore, possessed by this section are of great importance, and offer peculiar inducements to the settler. Its valley is not only capable of grazing immense bands of stock of every kind, but is also capable of supporting a dense population. The mountain slopes, on either side of the valley, and the land along the base of the mountains, afford at all seasons, even during the most severe winters, grazing ground in abundance, while the mountains are covered with a beautiful growth of pine. The provisions of nature here are, therefore, on no small scale and of no small importance; and let those who have imagined—and some have been so bold as to say it—that there exists only one immense bed of mountains from the head-waters of the Missouri to the Cascade range, turn their attention to this section, and let them contemplate its advantages and resources, and ask themselves, since these things exist can it be long before public attention shall be attracted to and fastened upon this heretofore unknown and neglected region? Can it be that we shall have so near our Pacific coast a section of country, of hundreds of thousands of acres, that will remain forever untilled, uncultivated, totally neglected? It cannot be.' "

How true the words of Lieut. Mullan, uttered but a few years ago; for, to-day, the Bitter Root valley boasts of its farms and settlements, as also the Hell Gate, with its tributaries, upon one of which (Deer Lodge) is situated Deer Lodge, or Cotton-wood City, a small town, built after the rural style of log and adobe houses. At this point the American Fur Company have an immense stock of goods, from whence they are disseminated to all their trading posts.

In these valleys much grain is already grown, and along the Bitter Root several flouring mills may be found. Produce brings a good price, and the increasing demand for bread-stuffs and vegetables at Bannock City, and other mining towns, will insure a more vigorous effort on the part of the husbandman. While the Bitter Root furnishes the mines with vegetables, the Hell Gate and Deer Lodge valleys are not lacking in the way of supplying the market with beef. Mr. John Grant, who resides in the Deer Lodge valley, has some two thousand head of cattle and one thousand horses and mules, all of which thrive, winter and summer; from the first of which, choice beeves may be selected at any time.

East of the mountains, *or rather North-eastern Idaho,* is traversed by the Missouri and its numerous tributaries, among which the most important is the Yellow Stone, whose source is high up in the mountains, from thence, winding its way eastward across the Territory, and flowing into the Missouri at Fort Union; thus crossing seven degrees of longitude, with many tributaries flowing in from the south, in whose valleys, in connection with that of the Yellow Stone, there are hundreds of thousands of acres of tillable land, to say nothing of the tributaries of the Missouri, among which are the Jefferson, Madison and Gallatin forks, along which settlements are springing up, and agriculture is becoming a lucrative business. These valleys, like those of the Hell Gate and Bitter Root, have a genial climate, and are inviting to the settler. They are surrounded with hills and mountains, clad with pine, while a growth of cotton-wood skirts the meandering streams that everywhere flow through them, affording abundance of water power.

The settlement located at the three forks of the Missouri is rapidly increasing, there having been over one hundred farms taken up during the past season. The first attempt at farming was made in the summer of 1863, which was a success, and indicates the productiveness of these valleys. Messrs. Wilson & Co. broke thirty acres last spring, planting twelve acres of potatoes, also corn, turnips, and a variety of garden sauce, all of which did well. The potatoes, as they informed me, yielded 200 bushels per acre, and sold at Virginia City, fifty miles distant, at 25 cents per pound, turnips at 20 cents,

John L. Campbell

GALLATIN CITY — THREE FORKS OF THE MISSOURI.

onions at 40 cents, cabbage at 60 cents, peas and beans at 50 cents per pound in the pod, and corn at two dollars per *dozen* ears. Vines of all kinds seem to flourish, and we see no reason why fruit may not be grown here, as the climate is much more mild than in many of the States where it is a staple.

The valley at the Three Forks, as also the valleys along the streams as they recede from the Junction, are spacious, and yield a spontaneous growth of herbage, upon which cattle fatten during winter. This valley, though unknown until recently to the eastern populace, had been selected by Brigham Young for the establishment of a Mormon colony, and had it not been for the gold discoveries in this region, would ere this have been occupied by the saints.

The Missouri and Yellow Stone rivers are navigable, the former as high as Fort Benton, for steamboats of ordinary capacity, and it is also navigable for light draught boats above the portage, at Fort Benton, as high up as the Three Forks. And we are credibly informed that parties in St. Louis are now constructing two boats of light draught for the upper river service, one to ply above the portage, and the other to ply between Fort Benton and the lower river boats, which may not—in stages of low water, as was the case the past season—reach their usual moorings at Fort Benton. Already a town has been located at the Three Forks, which bears the name of the east fork (Gallatin), and is rapidly becoming an important place. It was laid out in December, 1862, by a company from Bannock City, who wisely appropriated one-fourth of the lots to donation purposes, as follows: 75 lots to the first steamboat, and 50 to the second, that should arrive there; 50 to the first printing press, and 25 to the proprietor of the first stock of goods brought there; also, 50 to the first church built there; 7 to the first lady who should become a resident of the place, 6 to the next, and so on, down to one. All of the latter class of donees have received the fee, and many of the others. Situated at the head of navigation, in the heart of an agricultural country, surrounded as it is with rich mining camps, it does not require a deeply philosophical cast of mind to look down the vista of the future a few years only, to behold a beautiful city towering here, locked forever in the embrace of the surrounding hills.

The Yellow Stone river is also navigable for several hundred miles from its mouth, penetrating the heart of the agricultural and mineral region of

Eastern Idaho. And since such is the case, who shall say that her waters, heretofore unbroken by the steamer, will not soon swarm with the commerce of a new State, bearing away upon its bosom the products of enterprise and honest toil, thus adding another artery to the great commercial world, and opening new avenues, through which the long-hoarded wealth of the mountains may pour forth to bless mankind?

The monotony of this section is broken by the various chains or spurs of mountains coursing through it, or shooting off from the main chain of the Rocky mountains, among which are the Black hills, Wind River chain, Big Horn and Yellow Stone mountains, with many other spurs of less notoriety, which renders this section undulating, with ranges of mountains, clad with evergreens, between which are beautiful valleys and winding streams, where towns and cities will soon spring up to adorn these mountain retreats, and give room for expanding civilization.

Idaho and its Gold Fields

MINES AND SETTLEMENTS

In what we have already written, a foundation has been laid for the discussion of the principal attraction of this region,—its gold mines. These have already called from thirty to forty thousand settlers into Idaho, of whom two-thirds are west of the mountains; most of them coming from the Pacific coast, Oregon, California, and Washington territory. The lesser share, east of the mountains, are chiefly from the States; multitudes of them escaped from the States in rebellion, to avoid the troublous times that act has involved.

Any elaborate treatise upon the gold mines would require data not yet collected from any source. Even a list of the mines would only be accurate on the day on which it was written, since new diggings are continually being opened and explored. The gold is found under the same circumstances as in California, both in gulches and in quartz.

The mines on the western slope of the mountains, first attracted attention as rivaling California in their stores of wealth. The excitement created by these reports caused an immense migration thither; this, of course, led to new discoveries, sparsely spreading over much of the western portion of the Territory, and extending over upon the head-waters of the Missouri and Yellow Stone.

The Salmon River mines, situated upon the river of the same name, were the first to attract the gold-hunter. Florence City is the largest settlement in the Salmon River country, and the general depot for supplies. The Salmon River gold is rather of an inferior quality, being worth only from $13 to $15 to the ounce, while that of other mines in the Territory is worth from $16 to $19. The yield of these mines has been very large, and many fortunes have been made by those who reached them in time to secure good claims.

South of Salmon river is a large extent of country as yet wholly unexplored. On Clearwater river and its branches, north of Salmon, gold is found over a large extent of country, Elk City and Oro Fino being the principal centres of business and population. At the junction of the Clearwater with Snake river is situated the town of Lewiston, the principal capital of Idaho, which is the largest town in the Territory, up to which point Snake river is navigable for steamboats of light draught, thus making a continuous line of navigation from the mouth of the Columbia, with the exception of two portages on the latter river—one called "the Dalles," and the other "the Cascades," where the Columbia breaks through the range of mountains by that name. Upon the Borse river, extensive mines have recently been discovered, and the town of West Bannock built up. Mines have also been opened on the Flat-head river, and on the Big and Little Black Foot rivers, all of which are upon the west side of the mountains. On the east side, the mines are rich beyond calculation, the yield thus far having equaled the most productive locality of California of equal extent. The Bannock or Grasshopper mines were discovered in July, 1862, and are situated upon Grasshopper creek, which is a tributary of the Jefferson fork of the Missouri, three hundred and eighty-five miles north of Salt Lake City, and two hundred and eighty south of Fort Benton.

The mining district here extends five miles down the creek from Bannock City, which is situated at the head of the gulch, while upon either side of the creek, the mountains are intersected with gold-bearing quartz lodes, many of which have been found to be very rich. But little quartz mining has been done, from the fact that there are no mills in operation here as yet, except a small water-mill, which drives but four ill-constructed stamps. This mill is owned by Mr. Allen, formerly of McGregor, Iowa, and is crushing from the Dakota lode, having taken out many thousand dollars of a superior quality of quartz gold since June last, as will be seen by reference to a letter from that gentleman which we insert in another place. This lode is said to be one of the richest on the continent, and *yields from five hundred to three thousand dollars per cord.* There are many other lodes supposed to be equally rich, which have not been further prospected than to discover and stake them out. The gulch and bar mines here have yielded exceedingly well, and many fortunes have been made, and we question very much whether there ever have been any mines on the continent where claims have paid so generally as at this point, and at Virginia City mines, of which we shall speak hereafter. Claims that do not pay ten dollars per day to the hand are not worked, as the whole country seems auriferous and abounds in ounce diggings, and in many instances the sluices yield the shining metal by the pound. There are three ditches or flumes here bringing water into the mines; one is twelve miles long, and cost about fifty thousand dollars. Its volume is seven hundred inches, and sells at seventy-five cents per inch per day. The other two are of less capacity, but of equal importance, furnishing many claims with water. The bars—which are large slides or benches situated upon the sides of the mountains—are made available by means of water from these ditches, and pay exceedingly well. They are mined by tunneling under them and wheeling the "pay-dirt" out, when it is washed by sluicing. The pay-dirt of which we speak is that in which the gold is found; varying in depth, in some localities being but a few inches, while in others it is several feet deep; and is frequently found covered with many feet of earth, which must either be removed or the process of tunneling be resorted to. The pay-dirt is of a peculiar color, and is readily known by the experienced miner.

But few of the creek claims have as yet been developed, but such as have, pay enormously. Mr. Samuel Hackley, who is mining in the creek some two miles below Bannock City, has a very rich claim, as has also Mr. John Knowles. The former gentleman, upon the arrival of Chief Justice Edgerton, invited him to visit his claim, upon which occasion he presented the judge with a pan of dirt, from which his Honor with his own hand soon extricated FORTY-EIGHT DOLLARS OF SHINING DUST; this not coming up to the expectation of Mr. Hackley, another pan was presented, from which a sufficient amount was taken to swell the sum to seventy-five dollars. Mr. Knowles above named invited some ladies to visit his claim, when he followed the example of Mr. Hackley, and presented each of the ladies with a pan of earth which yielded from twelve to twenty dollars to the pan. It must be remembered that this dirt was taken from the bedrock, where the richest deposits lie, the gold, of course, gravitating thither; but these yields are not extraordinary; we have known as high as one hundred dollars taken from a single pan of dirt. These yields are no criterion, as earth that yields twenty-five cents to the pan, and even less, pays good wages, and where these large yields are got, frequently much labor has to be expended in the removal of surface earth, etc.

The claims mentioned above yield from two to ten hundred dollars per day, with a force of from four to eight men. We do not mention them as the only paying claims of the district, for there are many of equal richness, and when all the creek claims shall have been opened, its product will be immense. During last winter while the creek was frozen, many of the miners carried pay-dirt from their bar claims into their cabins, washing it with heated water, thus making from five to ten dollars per day; and Mr. Murry, who owns an interest in the Dakota lode, carried pay-dirt, or rather decomposed quartz, down the hill to his cabin in a flour sack, which yielded as high as one hundred and fifty dollars to the sack. Claims here sell all the way from one hundred to ten thousand dollars, according to their known richness, or the richness of the locality in which they are situated.

Bannock City is the largest village in this vicinity, and is practically a mining town. The houses are built of logs, and number about three hundred. It has about one thousand inhabitants; and one school in operation, but no church. It has stores, hotels and saloons, in the extravagant proportions common to a thriving mining settlement, and one exchange office. Among its merchants may be mentioned Messrs. George Chrisman, Lovejoy, McDonnel, and Carter & Co. It has a city organization, with courts, marshal and aldermen, and other officers.

Between this point and Salt Lake City Mr. Oliver & Co., run a weekly four-horse coach, carrying both mail and passengers. The fare through from Bannock to Salt Lake City is fifty dollars, the trip being made in about eight days. This line is well managed, and the coaches are commodious. Government has also established a post route between these two cities, which is to go into operation early the coming year, and a line of telegraph is also to be constructed between these two points next season.

Produce at Bannock was high when we left, which was in October, 1863. Flour was $25 per hundred; Bacon, 30 cents per lb; Hams, 60 cents; fresh Stakes, 15 to 25 cents; Potatoes, per lb, 25 cents; Cabbage, per lb, 60 cents; Coffee, 80 cents; Sugar, 60 cents; fresh Butter, $1.25; Hay, 10 cents per lb, or $30 per ton; Lumber, $150 per thousand. Wages ruled at $5 per day for miners and common laborers, and $6 to $8 for mechanics. Female labor ranged from $10 to $15 per week. Washing, from $3 to $6 by the dozen.

Bannock was partially depopulated during the past season in consequence of the rich discoveries made upon the Stinking Water, and the excitement consequent thereupon, but nevertheless it will become an important town as it is situated in a rich quartz district, which is even now attracting machinery, Col. Hunkins, of Galena, Illinois, having already located a steam quartz mill here, which will go into operation early the present winter. We were informed by this gentleman that there were two other mills of a similar character freighted up the Missouri the past season, which will reach Bannock early in the spring.

Thirty miles south-west of Bannock is situated Jeff Davis' gulch, or the Horse Creek mines. These mines were discovered in July, 1863, but owing to the scarcity of water in that locality, few claims only could be worked, but such as were, paid from ten to one hundred dollars per day to the hand. Mr. Baugh, of Omaha, as also several others, made snug fortunes by mining in this gulch the past season; and should there be much snow in the mountains the present winter, or rain in the coming spring, much gold will be taken out next summer, as the claims pay, from the roots of the grass down to the bed-rock, which saves stripping, and renders the gold more accessible to the miner. In these mines, good claims can be got by the first who reach there in the spring; many having been abandoned by the original pre-emptors in consequence of scarcity of water, and they are subject to re-pre-emption by any who may wish to occupy them.

Upon Grasshopper creek, twelve miles north-west of Bannock, discoveries were made in August last. Here are both surface and quartz diggings. The surface mines, however, were found to pay only moderately well, and are abandoned for the present. The quartz prospected well, but are left until such time as they can be developed by machinery.

It is an interesting fact for the investigation of the archeologist and historian, that in the mines last named were found abundant traces of ancient mining, in the shape of a "feed" and "tail" ditch, and a shaft where the miners had dug down to the gold. In this shaft is growing a pine tree one foot in diameter, attesting the period since the work was abandoned. Near by a couple of timber huts are still standing, in shape like an Indian wigwam, but bearing every mark of great age. In these regions dew never falls, and timber exposed to the weather lasts a long period.

It is believed that these ancient mines were the work of Spaniards, penetrating northward from Mexico in the last century, Mexican chronicles referring to some such expedition in quest of gold, which pushed far north, beyond the Great Desert.

And here, referring to the weather in this mountain region, the writer remembers seeing a simple grave of some emigrant who fell by the way-side, and whose resting place, months before, had been enclosed by a railing of split pine. The surface of the wood was still bright, as if just from the axe.

About the time of these discoveries, others were also made seventy miles north-east of Bannock, upon the tributaries of Stinking Water, a small stream that puts into the Jefferson Fork on the east side. Here several rich gulches have been discovered, that wind their way from the base of a high chain of mountains which runs parallel with the Madison Fork, between it and the Stinking Water.

These mines were first discovered by Mr. Fairweather, a gentleman from Minnesota; the first being made upon Alder creek, which, since its discovery, bears the name of Fairweather's gulch. This gulch is fifteen miles long and pays exceedingly well the entire length, the gold dust being very fine at the lower end and growing coarser as you ascend, until nothing but shot or coarse gold is found at its head. The large mountain from which the gold has been thrown—situated at the head of the gulch—presents indications of quartz, but owing to the richness of the surface-diggings, and the scarcity of men required to work them, quartz prospecting has been deferred until winter, when the great source of this immense wealth that lies scattered down the mountain gorge, will doubtless be developed. In this gulch are located Virginia, Nevada and Summit cities.

The yield of this gulch *alone,* has been immense since its discovery in July last, when people rushed in from all quarters, depopulating some of the surrounding camps. It is estimated by parties most able to judge, that the yield has been at least *half a million dollars* per week. Virginia City, situated in this gulch, is the largest town in Eastern Idaho. It is of but a few days' growth, yet it contains some five thousand inhabitants. It is built of hewn log houses, with the exception of a few stone buildings, and is indeed a business place.

There are two lines of coaches which run between this point and Bannock City. One is owned by Messrs. Oliver & Co., and the other by

Messrs. Peabody & Co. The fare is ten dollars. The mines here are unsurpassed in richness; not a claim has been opened that does not pay good wages, while many claims yield the precious ore by the pound. We have known as high as eight pounds to be taken out by one sluice in a day, with a force of from six to ten men. Mr. Barton, formerly of Ohio, took out $25,000 in sixty days; the largest yield in any one day was $1,874. We could mention several other parties whom we personally know to have made good fortunes the last season in this gulch, and very many have done well. We have no doubt but that an immense quantity of gold will be taken out at this point the next season.

Produce is higher here than at Bannock City. Flour was $30 per hundred, and other things in proportion. Lumber was $400 per thousand. The cause of its being so high was owing to its having to be sawed by hand, but as mills have been erected, of course it will sell at less figures next season. Wages were from five to eight dollars per day for miners, and mechanics got from seven to ten dollars per day. Female labor, at the boarding houses and hotels, was twenty-five dollars per week.

It should be remembered that green-backs, as currency, are not in circulation, and that these are "Dust prices."

Four miles south of this gulch is situated Brown's gulch, which is said to be very rich; few claims, however, have been opened as yet, it having been discovered late in the season. Fifteen miles north of Fairweather's gulch is located Beven's gulch, which is also rich, the diggings extending five miles along the creek. The gold found here is mostly coarse, and many claims pay "from the roots of the grass down," as the miners say. Here small fortunes are made quickly, the gold being more accessible than in most districts. This gulch heads at the base of the same mountain with that of Fairweather's, and already at its source rich quartz lodes have been discovered.

Three miles south of this gulch Harris' gulch is located, which is of recent discovery and equally rich. These gulches all head upon the same range, and run parallel with each other, emptying into the Stinking Water, as indicated upon our map.

There are also good mines upon the Balder creek, a stream putting into the Jefferson fork, on the north side, twenty miles above Gallatin. Near the head of this creek will be seen the head-waters of the Prickly Pear, which runs north-easterly, emptying into the Missouri. Upon this stream and its tributary, paying mines are known to exist, but the scarcity of water has thus far prevented their development to any great extent. Upon the head-waters of Big Hole, Deer Lodge and Salmon rivers, good mines have been discovered, which will soon yield up their treasures. The numerous branches of Snake river afford the shining color, and it is expected that new discoveries will attract the gold-seeker in that direction early the coming spring. Upon the Saskatchewan river (which flows northward from this Territory and empties into Lake Winnipeg) extensive placer or surface diggings have recently been discovered, which will also have their attraction the coming year. Quartz lodes are also known to exist in all the ranges of mountains in this country, and when enterprising men with capital shall penetrate these hidden regions, this section of Idaho will rival Colorado in her stores of auriferous wealth.

While gold has been found in paying quantities all along the rocky chain, its deposits are not confined to this locality, but sweep across the country eastward some hundreds of miles to the Big Horn mountains. The gold discoveries there cover a large area of country, but the prospecter can go only at the risk of his life.

I will quote from Mr. Vanderbert, who accompanied a prospecting party of fifteen from Bannock to that country early in the spring of 1863. He says:

> "We found the color in nearly all the streams putting into the Yellow Stone, and also in the tributaries of the Big Horn, but we were attacked by the Indians at night while encamped at the base of the Big Horn mountains, two of our men killed, and several others wounded, compelling us to fly for our lives, some of our party starting on our back track for Bannock, while I, with the rest, started for the Platte. We urged our steeds to their utmost during the succeeding nights, secluding ourselves in the gorges during the daytime. We soon reached the Platte in safety, and thus ended our prospecting tour at the Big Horn."

We quote the above to show our readers with what hazard of life the prospecter penetrates the dark recesses of our Western wilds in search of gold. But at this point, as at all others, the winding trail and advancing step of the white man will be irresistible, for wherever Mammon leads, there will also be found advancing enterprise, preparing the way for the ever onward march of civilization, before which the hostile Indian and the buffalo disappear forever.

Thus it will be here. Within a year the mountains of the Big Horn will teem with a mining population, and the Yellow Stone will smile a hearty welcome to her agricultural guests.

John L. Campbell

Idaho and its Gold Fields

THE INDIANS OF IDAHO.

The riches of this innermost shrine of the continent have not been untold in miners' tales, and have for years haunted the brain of the gold-seeker on the Pacific coast, who has long believed that *the* El Dorado was still to be opened. The fierce implacability of the Indian tribes of the region has maintained the barrier against the whites. These aborigines have made this mountain territory the last stronghold of their race on this continent, and farthest removed of all from the haunts of civilization they have maintained the ancient wilderness of the race that in earlier days fought the advance of the white man step by step crowding them from their domain.

The Indians of Idaho are the Snakes, Bannocks, Flat-heads, Black-foots, Nez Perces, and other smaller segments of once great, but now fading nationalities. Beyond all doubt hundreds of adventurous whites have paid the penalty of life for the fatal quest of gold in these mountain passes. Party after party has in time past been formed and pressed forward only to return scattered and broken from the fierce assault of the lurking savage. Referring to this subject, an Oregon paper of recent date says:

"A tradition has been current for years, that some lost immigrants, in 1845, while wandering through the country drained by the Malheur, discovered mines where gold could be raked up by the shovelfull. At the time, the discoverers were ignorant of the characteristics of gold in its native state, and accordingly they passed on, regarding the metal as worthless. A few years later, some of these men were attracted to California, and on visiting the mines there, almost the first remark was, that they knew where bushels of that kind of stuff were to be had. Since that date scarce a year has passed that did not witness the departure of companies of men who were sent for the purpose of discovering the country described by the immigrants. These exploring parties have uniformly proved failures, owing, in a great measure to the hostility of the Indians, who have resolutely refused to allow the white man to prospect their country. At last, however, a party more fortunate than the rest has succeeded in finding the long lost gold-field, and, if reports are to be believed, the story of its richness has not been exaggerated by the original discoverers."

There is no doubt that this long-locked region is now opened, and that Idaho will prove more marvellous than the ancient Ophir by its precious ores.

HYDRAULIC MINING.

DRIFT MINING NEAR BANNOCK CITY.

The large bars which overhang the creek in the Grasshopper diggings—of which we have already spoken in another place—are being washed by what is called the "hydraulic method," an improvement in the art of placer mining and washing, which originated in California, and which enables them to mine and wash ten tons of earth, where, under the old methods, they could scarcely wash one.

The process, so new to all but Californians, is well exhibited at Bannock City, and will be briefy described. The annexed engraving is from a sketch of one side of the claims, and shows their general appearance. On one side we see the bank or bluff, formed by the drift which has not been disturbed. The top of this bank is the general level of the surface or inclination, which reaches back to the base of the mountains, upon which there is an occasional pine tree. The frame, or staging, elevated above the surface, is a flume or open conduit for the water, and is highest in the back-ground, and the water flows towards the bluff, although in the picture the descent appears to be in the other direction. At the end of this sluice a long hose-pipe of leather is attached, and extends down along a favorable part of the bank, to the level of the bed-rock below. In the engraving the bed-rock is not seen, being completely covered by the large boulders that have been excavated from the bank, and washed.

The bank is not attacked by pick-axe and shovel, but a powerful jet of water, delivered through the hose from the sluice above, is thrown against its base. By this means the earth is soon washed away, and the overhanging mass of drift of earth and loose boulders falls to the ground. As rapidly as the finer portions are removed by the water, the loose stones and boulders are thrown back out of the way, while the smaller fragments, together with the sand, clay and gold, are carried by the water into a long drain or sluice-way, where the gold is collected. The operation is thus a continuous one, and the earth is not handled or transported except by the water.

The operation of *sluicing* is another striking and important feature in the art of mining, as practiced in these mines. Earth, gravel and stones are washed by hundreds of tons in a short space of time, without being handled.

The sluice is a long channel or race-way, to conduct the water or gravel, and is constructed either in the surface of the bed-rock by excavating, or made of boards. The former is known as the *ground sluice,* and the latter as the *board sluice.* A board sluice is generally twelve or fifteen inches in width, and from eight to ten inches deep, and is made in convenient lengths so that one can be added to another, until a length of two or three hundred feet or more is obtained. False bottoms of boards, pierced with holes, or a series of raised cleats, are placed in the bottom of the sluices, and are intended to receive and retain the gold, while the stones and gravel are washed away. Long bars, or a grating with the spaces parallel with the sluice, are, however, generally preferred to the cross-cleats or holes.

The fall, or rate of descent of the sluice is varied according to circumstances, being arranged to suit the size of the gold and the nature of the drift. One or two feet in a rod is a common inclination, and with a good supply of water, is sufficient to cause stones two or three inches in diameter to roll from one end of the sluice to the other. The board sluice is more commonly used where the earth has to be handled, which is accomplished by picking and shoveling. While ground-sluicing is a process of drifting or mining, where the earth is washed away by the introduction of a stream of water without a hose. The earth, stones and gold, as they enter these sluices with the water, are all mingled together, but the current soon effects a separation; the lighter portions are swept on in advance, and the gold remains behind and moves slowly forward until its drops down between the cleats or bars. The larger stones and coarse gravel are swept on by the current, and after traversing the whole length of the sluice, are thrown out at the lower end. The operation, as in the case of the hydraulic method, is a continuous one, and requires little labor or attendance, except to keep the sluices from clogging. This is done by one or two men who walk up and down and throw out the larger stones with forks. The water for these operations, at such a height above the creek, and for the elevated placers or "dry diggins" generally, is brought in aqueducts or ditches from the sources of the streams, many miles distant, in the mountains.

The water at Bannock City is supplied by the following named companies, before alluded to the *West Ditch Company,* the *North-west Ditch*

Company, and the *North Side Ditch Company.* The water, after traversing these ditches for many miles, is sold to the miners, being delivered from horizontal apertures, the openings of which are graduated to half inches; and for each inch of water the miners pays from fifty to seventy-five cents, for each day of ten hours.

John L. Campbell

Idaho and its Gold Fields

RIVERS

By a glance at a map comprising this new Territory, it will be seen that many of the great rivers of North America have their source within it. The Missouri river not only rises in Idaho, but winds its way for a distance of seven or eight hundred miles across the Territory. The Yellow Stone is wholly within its limits.

The North Platte also rises in this Territory upon the east side of the mountains, and flows easterly crossing the bounds of the Territory some thirty miles east of Fort Laramie. Upon the western slope of the Rocky mountains, and within this Territory, the great Rio Colorado rises, which flows into the Gulf of California. The Salmon river, whose mines have created such an excitement and proved to be exceedingly rich is wholly within its border.

Both the great tributaries of the Columbia—Snake river and Clark's fork—have their sources in Western Idaho. The Hell Gate, Bitter Root and Kooskooskee all originate in Idaho, and flow westward into the Pacific.

The Saskatchewan river, which flows northward across the British Possessions, mingling its waters with those of Lake Winnipeg, also has its source amid the mountains of this Territory.

Idaho is indeed the great fountain-head of the continent, from whence the waters diverge to the four points of the compass: easterly, emptying southerly into the Gulf of Mexico; southwesterly, into the Gulf of California; westerly, into the Pacific ocean; and northerly, into Lake Winnipeg, thence into Hudson Bay.

It may well be said of Idaho that hers are the springs of the great water highways of the continent, as also hers the source of incalculable riches, which will hereafter be derived from her extensive mines to swell the wealth of nations.

As usual in new gold regions, gulch mining, as the easiest worked, is first resorted to, but this is mere gleaning as compared with the solid and permanent yield of quartz mining when properly carried on. A private letter from Bannock City, an extract from which we insert below, gives details of a simple and effective quartz mill introduced there by Mr. Allan, being the only one as yet in operation in Eastern Idaho. Others more complicated and costly are being introduced. We give above a diagram of Mr. Allan's apparatus.

"My first operation was to put up a very rude structure in the shape of a quartz mill formed in this way. An over-shot wheel, twenty feet in diameter, is placed on a shaft eighteen feet long, with large pins in the shaft for the purpose of raising the stamps. These stamps are fourteen feet long and eight inches square, and strapped with tire iron on the bottom, which work into a box that is lined on the sides with copper plate galvanized with quicksilver, so as to catch the gold as the quartz is crushed and dashed up on the sides of the box. Then we have an opening on one side of the box, with a fine screen in it through which the fine quartz and fine gold pass, and run over a table covered with copper. The quartz lodes here are very rich. We have taken out between fifty and sixty thousand dollars from one claim alone. We have five claims. We have picked up pieces that weighed two and three ounces of beautiful gold.

J. F. Allan."

Idaho and its Gold Fields

WHAT IS NECESSARY FOR AN OUTFIT.

Let us suppose that a party of men have arrived at the Missouri river who are going to Idaho, and wish to prepare for a trip across the plains. What kind of a team and wagon is most advisable for a trip across the plains. What kind of a team and wagon is most advisable to take? What variety of provisions will be most suited to a journey of this kind; what quantity ought a party of four men to take; and how should it be prepared? What mining and other implements is it necessary to provide? Will the party need a tent when their wagons are covered? What arms, if any, should be carried, and what supply of clothing should a person have for a trip of this kind?

The first question to be settled is, what kind of a team is it advisable to take—should it be horses, mules or oxen? For many reasons the ox is preferable. Firstly, a team of this kind is much cheaper than either of the others. Three yoke, at $75 per yoke, would amount to $225, and would be fully equal to two span of horses or mules, which would cost double that sum. Secondly, they require less feed and attention, and very seldom stray so as not to be readily found, neither are they as liable to be stolen, or stampeded by the Indians, as horses or mules. As to a wagon, it does not require an expensive one; just such a one as a farmer would select to do his farm work (a common lumber wagon) is the most suitable. This kind will meet with a ready sale in the mines, whereas more expensive wagons with springs and stationary covers are in less demand. It should be made of the best of seasoned lumber and put together firmly so as to stand the drouth of the plains. The thimble skein axle is preferable. It should be covered with canvass, and it would be well to have it lined overhead with oil-cloth so that goods will be protected from the weather, however hard it may storm.

As to provisions and the variety suited to take, first we say that no party should leave the Missouri river next spring for Idaho, without a supply sufficient TO LAST THEM NINE MONTHS. The emigrant may ask, why cumber our wagons with such an amount when we shall be but sixty or seventy days on the route? But remember, you are not going to an agricultural country, or at least one developed, but are going to a very new section where produce is scarce and high, and has to be freighted many hundred miles; and should all go with just enough to last them through, much suffering would be inevitable, and more particularly so the coming season, from the fact that the surplus of produce grown in Utah the past season has already been freighted to this new Territory, and bears a high price, owing to the large emigration that has already reached the mines. When we contemplate the immense emigration that must inevitably pour into that country from both the East and the West the coming season, we can but admonish all who go to be sure and carry provisions enough to last until after another crop shall have been grown in Utah. There are further reasons why parties should take a good supply. It may be some little time after arriving before getting into business, and to have to pay twenty-five or thirty dollars per hundred for flour, and for other necessaries in proportion, or fourteen dollars per week for board, would be too great a drain upon the pockets of many; hence go prepared.

Each party should take at least one good cow for milking purposes, as milk is relished upon the plains, and on many occasions when great dispatch is required in the getting up of a meal, or in case of a storm when cooking cannot be done, it is resorted to, and serves a tolerable purpose. A tent too is almost an indispensable article, and each party, however well their wagons may be covered, will do well to take one. They are readily pitched, and with a stove situated in one corner with the pipe protruding through the roof, they answer the purpose of a house, and with a good supply of catables, one can be "quite at home," however distant from civilization; whereas, without one, the party must cook in the open air whatever the weather may be, and the sleeping apartment in the wagon, too, is not desirable, from the fact that it is always stored with boxes, kegs, etc., while the tent furnishes a comfortable sleeping place, which is one of the requisites to health in a trip of this kind.

As to a stove, many, indeed nearly all who cross the plains, use what is called the "emigrant stove," which is simply a small sheet-iron stove answering a very good purpose, but which soon burns out, frequently not lasting through the trip. The common cast-iron cooking stove, which sells in the States for from eighteen to twenty-five dollars, sells in the mines for from one to two hundred dollars, and may be readily carried and used on the way, and upon arriving if desired, it will sell for at least one hundred dollars profit, whereas the sheet iron stove will be comparatively worthless.

Idaho and its Gold Fields

In regard to clothing, persons had better be too warm than suffer from cold, yet it will not be necessary to take a very large amount; say one or two extra suits of good durable clothing are sufficient. Each person should have a rubber coat and leggings, also two pair of woolen blankets or similar bedding of some kind. The emigrant should have two extra pair of double-soled boots. Parties should go well armed. Each should have a rifle or shot gun, and a revolver. Very few who cross the plains have occasion to use them, but the fact of having them along serves to fortify parties against an attack from either the marauding whites or hostile Indians.

A pony is not an indispensable requisite to a trip of this kind, yet it is advisable for a party to have one along; they can be had cheap at the Missouri river, and will save many a step for the weary emigrant in the way of herding and collecting his stock; and for the purpose of enjoying the buffalo chase or the more daring encounter of the grizzly, the pony is quite indispensable.

The following table comprises the necessaries for an outfit of nine months for four persons:

3 yoke of Oxen, $75 per yoke	$225.00
1 Wagon and Cover	100.00
1 Tent	15.00
12 sacks of Flour	36.00
400 pounds of Bacon	40.00
100 pounds of Coffee	30.00
40 pounds of Candles	10.00
10 pounds of Tea	10.00
Yeast Powders	5.00
50 pounds of Salt	1.00
3 pounds of Pepper	.50
2 bushels of Beans	3.00
15 gallons of Vinegar	4.00
25 pounds of Bar Soap	3.00
50 pounds of Lard	5.00
1 gross Matches	1.00
1 ten-gallon Water Keg	1.25
1 Coffee Mill	.75
2 Coffee Pots	1.50
8 Tin Plates	.50
8 Tin Cups	.50
2 Frying Pans	1.00
4 Butcher Knives	2.00
6 Knives and 6 Forks	1.50
200 pounds of Sugar	25.00
Carried forward	$522.50

Brought forward	$522.50
1 Skillet	1.50
2 Water Buckets	.50
2 small Tin Pails	1.00
75 feet of Rope	2.50
6 Table Spoons	.50
2 Camp Kettles	1.25
4 Gold Pans	3.00
4 Picks	5.00
4 Shovels	5.00
2 Axes	2.50
2 Bread Pans	1.00
1 Wagon Bucket	1.00
Hand Saw and Drawing Knife	2.00
2 Chisels and Augers	2.00
1 Pair of Gold Scales	4.00
2 Files	.50
Hatchet and Hammer	1.00
2 Gimlets	.25
10 pounds of cut and Wrought Nails	.75
1 Whetstone	.10
4 bushels of Dried Apples	6.00
1 bushel of Dried Peaches	2.00
50 pounds of Rice	5.00
Total cost	$570.85

Should the party take a cooking stove with furniture, many of the above mentioned articles could be dispensed with, by an economy of use.

As to diet suited to the plains, very many who cross the plains seem to think that none of the luxuries of home can be enjoyed in a trip of this kind. From this fact they provide themselves with only breadstuffs and meats, while fruits, butter and eggs are left quite out of the bill. We have observed a very great difference as to the health of parties. Those who use meats and little or no fruit, incline to the scurvy, while those who use fruits and very little bacon or meat, never have it. Bacon and hams should be snugly packed in a wagon where the sun cannot reach them, nor should they be frequently spread upon the ground in the sun, as is often the case, as they will soon taint, but should be kept dry and seldom moved. Fruits, either canned or dried may be carried with perfect safety, and a good supply of the latter should be taken. Butter too may be carried in safety by putting it up in cans. From ten to fifty pounds may be put into a can, and it will be highly relished, and should be taken by all means. Eggs packed in a box with oats or bran may be carried for use during the trip. The emigrant will find that these articles will add much to the luxuriousness of his table, and render camp life more like home.

Idaho and its Gold Fields

THE REALITY.

Have a good reason for breaking the old moorings before looking for better ones, and when you start on a trip of this kind, do not cherish the idea that it is to be but a holiday excursion, soon to be over, when you will tumble into some rich gulch, only to come forth laden with stores of gold.

To succeed in any new field of labor, great industry and perseverance is required, and the emigrant to Idaho will secure his fortune only through hardship, privation, endurance and great industry. *Let well enough alone* when you are comfortably situated, and do not believe every story that goes the rounds. Few who appreciate these facts and go fully determined will fail to prosper.

WHO SHOULD NOT GO.

Persons who have good homes and means of livelihood, should not be induced by extravagant stories, however true they may be, to emigrate to a far-off country after a phantom fortune. Neither should any man who is so indisposed to labor as to have always failed at home to obtain an honest living, ever think of succeeding in a mining country, however rich it may be.

PROFITABLE PURSUITS BESIDES MINING.

All cannot mine. Some must make shoes, some follow blacksmithing, others work in wood; and the choice farming lands adjacent to every mining camp will be immediately put under the most profitable cultivation. Simply digging gold or other precious metals is a lottery in which there are many prizes but very many blanks; and I doubt whether there is a class of people in the world who succeed generally so well in life as the mechanic and the industrious farmer, especially when these vocations are followed in the vicinity of productive mines.

John L. Campbell

HOW TO GO TO IDAHO

DESCRIPTION OF OUTFITTING POINTS.

FROM ST. JOSEPH, MISSOURI.

St. Joseph is the largest city above St. Louis on the Missouri river. It contains about fifteen thousand inhabitants. It is beautifully situated upon the east side of the river, at the terminus of the Hannibal and St. Joseph Railroad, which thus connecting it with the great railroad system of the country, has given it prominence as an outfitting point, and has secured to it not only a substantial growth, but also an immense outfitting business. It is surrounded by a fine farming country which has long made it the mart of teeming products.

In this city are situated some of the heaviest outfitting stores of the frontier. From this point an immense emigration leaves annually for Colorado, California, Idaho and other places of the far West. A good steam ferry plies between St. Joseph and Elwood, a small town situated upon the Kansas shore.

The road leading from St. Joseph to Fort Kearney passes through a settled country, and the experience of the emigrant on this part of the route, will be much the same as it would be in crossing any of the Western States, as numerous settlements are to be found. Along this route Ranches abound, with abundance of feed and water.

A party going west from St. Joseph would do well to make up their full outfit there. The claims of St. Joseph as a post for outfitting are enforced by the names and reputations of its numerous business houses, cards of which may be seen in the following pages.

John L. Campbell

TABLE OF DISTANCES FROM ST. JOSEPH TO FORT KEARNEY.

St. Joseph to

Kinnekuk	40
Kickapoo	12—52
Log Chain	13—65
Seneca	11—76
Laramie Creek	12—88
Guittard's	12—100
Okaw	10—110
Otoe	11—121
Pawnee	11—132
Grayson's	14—146
Big Sandy	10—156
Thompson's	14—170
Kiowa	14—184
Little Blue	12—196
Liberty Farm	13—209
Lone Tree	15—224
Thirty-two Mile Creek	10—234
Summit	12—246
Hooks	12—259
Fort Kearney	10—269

Idaho and its Gold Fields

FROM OMAHA AND COUNCIL BLUFFS.

Omaha, the capital of Nebraska, is the largest city in that rapidly growing Territory. It is directly opposite the lively city of Council Bluffs, the leading place in Western Iowa, though the lesser of these two cities. A good steam ferry connects them. A steamboat line runs between the Hannibal and St. Joseph Railroad terminus and this place. Omaha is fixed by the President of the United States as the point of departure for the Union Pacific Railroad, the ground for which great national enterprise was broken a short time since, with imposing ceremonies. Omaha is surrounded by a fine farming country. As a business point it has prominent claims for parties making up their outfit for the mines. Our advertisements furnish a reliable key to the attractions of this character, both in Council Bluffs and Omaha City. The great through route from Omaha to the gold regions of Idaho, is the one we have chosen for detailed notes and tables of distance. Let us first speak of other points of departure on the Missouri river, and of Fort Kearney, in the Platte Valley, not far from which the routes from these various points concentrate.

Omaha is destined to be the market and entrepot of the Platte Valley, through which is to pour the tide of emigration and travel, swelling each year and extending its facilities, until the slow train of the emigrant gives place to the rushing railway train, with tireless steeds and iron pathway.

John L. Campbell

FROM NEBRASKA CITY.

Nebraska City, N.T., on the Missouri river about forty miles below Omaha, is one of the most thriving towns west of the Missouri. It already numbers 3,000 inhabitants, and a fine country all about it is making it the market of teeming products. About thirty miles west, a collection of salt springs have attracted capital, and the present salt works will be made more extensive, and the product promises to be unlimited. At Nebraska City, the party going west from there, would do well to make up their full outfit. From this point some of the heaviest freighting trains leave annually for various points in the mountains. For the share of the route as far as Fort Kearney the emigrant's experiences are not unlike what they would be in crossing any of our newer Western States, for settlements are springing up all along the route, and good ranches abound, with abundant feed and water. But outfits cannot be so well procured after leaving the Missouri river. The cards we give of Nebraska City firms, are a very reliable business directory for the emigrant. It should be added that the traveler will find a new tri-weekly line of excellent stages, just put on, running to Fort Kearney, making the trip in forty hours. Fare about $15. The following is a

TABLE OF DISTANCES FROM NEBRASKA CITY TO FORT KEARNEY.

Nebraska City to

Wilson's Bridge	9
Brown's Bridge	7—16
Bridge over Little Nemaha	4—20
Head of Little Nemaha	20—40
Olatha	8—48
Beranger	18—66
Grove of timber	15—81
Head of north fork Blue river	48—129
A pond side of the road	5—134
Prairie Lake	8—142
Junction of roads	5—147
Fort Kearney	30—177

Idaho and its Gold Fields

FROM ATCHISON, KANSAS.

Atchison is a town on the west bank of the Missouri, twenty miles above Leavenworth. It has a thriving appearance, and for two or three seasons past, has been the point of departure of much emigration. Holliday's line of stages for Salt Lake, by way of Denver, leave this place daily. Time to Salt Lake, 12 days; fare, $150. The traveler will find these coaches commodious, with the most obliging drivers and station keepers. By this line the mines of Idaho may be reached in twenty-five days less time than by any other route. One dollar and a half a day will be a safe average for meals.

John L. Campbell

Idaho and its Gold Fields

FROM LEAVENWORTH, KANSAS.

Leavenworth is a thriving city of from 10,000 to 12,000 inhabitants, on the site of the old military post of the same name. It is the largest place west of the Missouri, and, from location, bids fair to maintain its prominence as the centre of an immense outfitting trade. It is the principal point of departure for emigration from Missouri and the Southern States. The number of refugees from the troublous times in the South is immense, and only to be counted by thousands.

Congress authorized, last winter, the building by the Leavenworth and Pawnee Railroad Company, of a road from the mouth of the Kansas Valley, up the same westward to the point at the 100th meridian west, where the Union Pacific Road unites with the Iowa and Nebraska Roads. The Leavenworth Company had already been endowed with lands from the Delaware and Potawatomic Indian Reservations. They receive, in addition, the $16,000 per mile, of bonds granted by Congress, and were required to connect at the mouth of the Kansas with the Missouri Pacific Railroad, as also to build a road from Leavenworth across the Delaware lands to a point at or near Lawrence, making a junction with the Valley Road. Last Spring the original company sold out to a new one, the President of which is General John C. Fremont. This company contracted with Samuel Hallet & Co., of New York, to build the road. It is being done with almost marvellous energy. About thirty miles are graded, from the mouth of the Kansas, and the grading of the Leavenworth Road is nearly completed.

The following is a

TABLE OF DISTANCES FROM LEAVENWORTH TO FORT KEARNEY.

Leavenworth to	
Lancaster	30
Huron	13—43
Kinnekuk	10—53
Walut Creek	8—61
Locklands	13—74
Minnehah	18—92
Ash Point	12—104
Black Vermillion	12—116
Elm Creek	10—126
Big Blue, Marysville	14—140
Cottonwood Ranche	12—152
Rock Creek	20—172
Little Sandy	14—186
Big Sandy	5—191
Creek	8—199
Little Blue River	9—208
Leaving of Little Blue	44—252
32 Mile Creek	8—260
Sand Hill Pond	14—274
Platte River	8—282
Fort Kearney	12—294

John L. Campbell

Idaho and its Gold Fields

FROM FORT KEARNEY, NEBRASKA TERRITORY.

Fort Kearney is a long established military post, bearing the name of a well-known Federal officer, prominently associated with the military history of the Far West. It is 197 miles from Omaha. Several of the routes above named, starting from points south of Omaha, concentrate here; and there is thence westward a choice of two routes, on the north or the south side of the Platte river. There is nothing that resembles military defensive works here, it being simply a military rendezvous, with barracks and parade ground. The force now here is the 7th Iowa cavalry. A mile above, on the Platte river, the little town of Kearney City has three hundred inhabitants, and is slowly growing as a trading post. The Platte river is one of the main tributaries of the Missouri, into which it empties at Platte's mouth, a few miles below Omaha. Throughout its whole length the Platte is a shallow stream, straying through shifting, sandy channels, and among numerous islands. One of these, Grand Island, is sixty miles long, extending from a little below Fort Kearney. At the Fort the river bed is two miles wide. The characteristics of the Platte valley are attracting great attention among those who have the soberer quest for permanent homes. Its agricultural wealth is equal to the best portions of Illinois and Missouri. Already farming has been made unusually profitable by the ready market for cereals furnished by passing trains of emigrants. In the spring of 1863, corn was selling for one dollar per bushel. Through this fertile and beautiful valley, the Union Pacific Railroad, already begun at Omaha, is to run. The timber of this section is mostly cottonwood; along the river and on the bluffs, and smaller streams back from the Platte, oak and hard woods are found, but only thinly.

The towns in this valley are Columbus, at the mouth of Loup Fork, the county seat of Platte county, a pretty little place of 300 inhabitants, Fremont, about forty miles further down the Platte, the county seat of Dodge county, is an older town than the last named, but has less life. Twenty miles west of Omaha, on the Elkhorn river, a tributary of the Platte, is Elk City, a small village. All the above named are noted, with distances, in the table following.

The Western Stage Company run a four-horse coach from Omaha to Fort Kearney, leaving every other day, making the trip of 197 miles in about forty hours. The fare is $15. Good places for beds and meals are to be found by the traveler, on the route. The Overland Telegraph has offices at these various points. The roads are fine, with no sharp ascents. The soil tracks well, and is hard and smooth three-fourths of the year. At Fort Kearney the emigrant bids adieu to civilization, and thence onward must rely upon his own resources, striking out boldly for new regions. He may indeed become a passenger of Ben Holliday's stages for Salt Lake, a trip of nine days from Fort Kearney, at a fare of about $135. The coaches are serviceable and commodious enough, and eating stations are found at required intervals along the route. The emigrant party leaving Fort Kearney may cross the Platte, or follow the longer route by the south bank. Both routes are taken.

THE SOUTH ROUTE FROM FORT KEARNEY.

FROM FORT KEARNEY TO SALT LAKE *via* DENVER CITY.

TABLE OF DISTANCES.

Omaha to	
Fort Kearney	197
Platte Station	10—207
Craig	11—218
Plum Creek	15—233
Willow Island	15—248
Midway	14—252
Gilman's	15—267
Cottonwood Springs	17—284
Cold Springs	15—299
Fremont Springs	14—313
Elkhorn	11—324
Alkali Lake	14—338
Sand Hill	12—350
Diamond Springs	11—361
South Platte	15—376
Julesburg	14—390
Antelope	12—402
Spring Hill	13—415
Dennison's	13—428
Valley Station	12—440
Kelly's	15—455
Beaver Creek	12—467
Bijou	20—487
Fremont's Orchard	16—503
Eagle's Nest	11—514
Latham	12—526
Big Bend	15—541
Fort Lupton	17—558
Pierson's	15—573
Denver	14—587
Child's	11—598
Boon's	12—610
Little Thompson	18—628
Big Thompson	8—636
Laporte	16—652
Boner	10—662
Cherokee	12—674
Virginia Dale	12—686

Idaho and its Gold Fields

Willow Springs	15—701
Big Laramie	15—716
Little Laramie	14—730
Cooper Creek	17—747
Rock Creek	11—758
Medicine Bow	17—775
Elk Mountain	8—783
Pass Creek	14—797
North Platte	16—813
Sage Creek	14—827
Pine Grove	10—837
Bridger's Pass	9—846
Sulphur Springs	10—856
Waskie	11—867
Duck Lake	13—880
Dug Springs	12—892
Laclede	15—907
Big Pond	12—919
Black Buttes	14—933
Rock Point	14—947
Salt Wells	14—961
Rock Spring	14—975
Green River	15—990
Lone Tree	15—1005
Ham's Fork	18—1023
Church Buttes	12—1035
Millersville	8—1043
Fort Bridger	13—1056
Muddy	12—1068
Quaking Asp Springs	10—1078
Bear River	20—1098
Needle Rock	10—1108
Echo Canyon	10—1118
Hanging Rock	10—1128
Weber	10—1138
Daniel's	12—1150
Kimball's	11—1161
Mountain Dell	15—1176
Great Salt Lake City	14—1190

SALT LAKE TO BANNOCK CITY.

Salt Lake City to

Farmington	17
Ogden	20—37
Brewery	7—44
Brigham City	15—59
Box Elder	4—63
Wellsville	11—74
Logan	9—83
Summit Creek	8—91
Franklin	12—103
Bear River ferry	10—113
Deep Canon	14—127
Cotton-wood Creek	6—133
Bear River	10—143
Canon Creek	2—145
Soda Springs	16—161
Black Foot Creek	12—173
Black Foot crossing	17—190

Here the route intersects with the North Platte route at Independence Valley, shown in the extended table above.

THE EMIGRANT'S GUIDE

FROM THE MISSOURI RIVER TO IDAHO.

(NOTE.—The following, it should be stated, does not give the full list of all the ranches on the route, but those given are sufficient and reliable, and the success of the emigrant in finding others will still give him no cause to complain of misrepresentation in respect to accommodations on the route.)

One mile from Omaha, you will find good accommodations for your stock	1
Little Papillon	8— 9
Papillon	4— 13
Reed's Ranche. Water, grass and good camping	3— 16
J. F. Munger. General accommodations	3— 19
Elkhorn City. Good accommodations for emigrants	3— 22
Bridgeport. (On Elkhorn river, one mile from Elkhorn City. Several stores and large settlement,—good camping ground for the night. Wood, water and grass.)	1— 23
Farmers' House. Good accommodations; plenty of water and grass..	11— 34
Fremont. Small town and settlement	3— 37
Dale House. Corn meal, hay and stabling. Good camping ground....	3— 40
North Bend. A good camping ground here. The Platte strikes the road.	12— 52
Ranche and store	2— 54
Platte Valley House: by R. Graham. Blacksmithing, wood, water and grass	1— 55
Buchanan House, at Shell Creek. Wood, water and grass	8— 63
Sixty-nine Mile House—from Omaha. General accommodations; good water and grass	6— 69
Junction Ranche: by H. Bushnell. General accommodations	2— 71
Joseph Russell's. Wood, water and grass	1— 72
Peter Murie's. All kinds of produce for sale. Good camping ground.	10— 82
Columbus. Situated on the north branch of the Loup Fork. Ferry across here. The last town you will pass. Here secure any needed supplies not before secured	3— 85

Crossing Loup Fork, the next ranche is

Guy C. Barnum's. Good camping ground	1— 86
Prairie Creek Ranche. Good accommodations. Creek is bridged....	11— 97
James Cummins' Station	9—106
Lone Tree Ranche. Groceries, hay, corn and stabling. On the bank of the Platte.	25—131
Station, by Samuel G. Hayward. Good camping ground	1—132
E. D. Hurley's. Groceries, stock of all kinds kept	10—142
Jesse Shoemaker's Point. Good accommodations	1—143
Grand Island City	10—153
Wood River	10—163
Boyd Brothers. Nebraska Centre post office. Brewery and blacksmith shop	22—185
Miller & Co.'s Ranche. (Opposite Fort Kearney.) Hay, corn, stabling and general accommodations. On the bank of the Platte, at the crossing. Here the river is divided by several islands, and is two miles in width; difficult crossing at high water	10—195
Deep ravine. Steep descent	13—208

Two and a quarter miles beyond, is a good place for camping, on a low bench, twenty rods south of the road. No timber but willow.

Deep Dry Creek. No timber on it	3—211
Deep Dry Creek	6—217

Head of Grand Island is about opposite to this creek.

Elm Creek. Deep bank; plenty of timber	4—221
Road leaves the river near timber. Good camping place	6—227
Crossing of Buffalo Creek	4—231
Road runs near the river. Grass, wood and water	13—244
Willow Lake. South of the road. Good place to camp, but no timber	7—251
Utah Lake. South of the road	8—259
Deep Dry Creek	2—261
Low sandy bluffs, extending to the river	14—275

You will not strike the river for sixteen miles, but will have no difficulty in finding feed and water.

Skunk Creek. Six feet wide	5—280
Crossing of Skunk Creek. No timber, but grass and water	6—286
Good spring of cold water. At the foot of the bluffs, north of the road, at the head of the Pawnee Swamps	7—293
Low, sandy bluffs. Opposite the junction of the North and South Platte. Altitude, 2,685 feet	1—294
Carrion Creek. Ten feet wide and one foot deep. Good place for grass but no timber	3—297
Road, river and timber. Good place to camp	5—302
Wide, deep creek. Here the party can get Willow-brush for fuel	6—308
Black Mud Creek. Little feed for teams	3—311
Small Creek. Steep banks; very little water	6—317
North Bluff Fork. Six rods wide, and two feet deep	3—320
Sandy Bluffs. West foot	6—326
Sandy Bluffs. East foot	4—330
Bluff Creek. Four feet wide and one foot deep	2—332
Small Creek,—running between the bluffs	6—338
Bluff spring, and small creek	3—341

In the neighborhood of these creeks the land is swampy and soft.

Goose Creek. Thirty feet wide, and three inches deep	3—344

Low range of bluffs, sandy, one-fourth mile wide. Many springs of cold water at the foot of the bluffs.

Small Creek. Four feet wide	2—346
Shoal Stream. Three feet wide	2—348
Rattlesnake Creek. Twenty feet wide and eighteen inches deep	4—352
Creek. Six feet wide. Land sandy	7—359
Camp Creek. Eight feet wide	4—363

Two creeks here, about the same size, but a few rods apart.

Creek. Three feet wide	4—367
Wolf Creek. Twenty feet wide	2—369

At the east foot of sandy bluffs you will probably have to double teams, if heavily loaded.

Watch Creek. Eight feet wide, and two feet deep	4—373
Ash Hollow. South side of the river	7—380
Castle Creek. Six rods wide, and two feet deep. Swift current, water muddy	3—383
Castle Bluffs. South side of the river	5—388

The party will cross but few creeks of water for twenty-five miles.

Sand Hill Creek. South side of the road	1
Creek, or Slough	9—397
Dry Creek. Thirty feet wide.	5—402
Crab Creek. Twenty feet wide; very shoal	4—406

Two miles further you will find some high bluffs on the right. By ascending one of the highest, the visitor will see Chimney Rock to the west.

Small lake, south of road	2—408
Cobb's Hills, west fort	8—416

Party will find it sandy for ten miles.

Ancient Bluff runs north side of road, resembling the ruins of ancient castles	2—418
Road joins the river,—good place to camp	11—429
Low, sandy bluff, west fort	9—438

The land for several miles is soft in wet weather, but good traveling in dry weather.

Chimney Rock, south side of river	14—452

Here the land begins to be sandy and barren. Prickly pears and wild sage continue during the remainder of the journey.

Scott's Bluffs, south side of river	20—472
Spring Creek, south side of road	4—476
Road runs near the river,—good chance to camp	12—488
Creek 200 yards south of road	5—493

By ascending the highest bluff, a view can be got of Laramie Peak, in the Black Hills.

Timber north side of the river	11—504
Raw Hide creek. Plenty of grass	6—510
River opposite Fort Laramie	12—522

FORT LARAMIE consists of both military and trading station. A good assortment of merchandise is kept here.

Dry Creek	4—526
Good cold springs on right of road, near cottonwood trees	7—533
Road joins river. Wood, water and grass	7—540
Alder clump on left of road. Good place to camp	9—549
For the next nine miles the road is somewhat sandy	9—558
River. Good camping ground; wood, water and timber	9—567
The road is now rough. After leaving the river four miles, the road descends to the river again	4—571
Steep and craggy ascent. Road mountainous	10—581
Road descends near the river. Sandy	5—586
High, rolling, barren country for some distance. Low lands bordering upon the river	16—602
Ferry of the Platte. Road rough	3—605
Upper Platte ferry and ford	2—607

Plenty of feed and some timber.

Road turns south and rises a long hill	7—614
Mineral springs and lake. No bad taste to the water	6—620
Rock avenue and deep descent	7—627

The road here passes between high rocks, forming a kind of avenue or gate way, for a quarter of a mile.

Alkali swamps and springs. The party must avoid camping here. There is a creek north-west, in timber, and better grass	2—629

Small stream of clear spring water. Good camping place	4—633
Plenty of grass; no wood, but plenty of buffalo chips.	
Prospect Hill. Pleasant view of the surrounding country to the Sweet Water mountains. Antelope and deer..........................	4—637
Bad slough. Plenty of grass, but little water. Hilly	4—641
Small creek left of road. Grass plenty. Use buffalo chips..........	6—647
Grease-wood creek, six feet wide, one foot deep. Very little grass, no fuel but wild sage. Road to Sweet Water sandy.................	2—649
Alkali springs and lakes. Here gather saleratus from a lake west of the road. The land is swampy and smells bad...................	6—655
Sweet Water river, eight rods wide, two feet deep, swift current. Grass plenty, but little wood..	4—659
Independence Rock and road. On the north side of the river, about 600 yards long and 120 wide, composed of hard granite. Here a party of emigrants celebrated the Fourth of July, and named the rock..	1—660
Devil's Gate. A little west from the road	5—665
The river here passes between perpendicular rocks 400 feet high. Fremont passed through the gate in a skiff.	
Creek, six feet wide. Good camping ground..........................	1—666
You will find grass all along the river, but no timber.	
Deep ravine and creek..	6—672
Road leaves the river and passes over a high bluff..................	4—676
Sage Creek. No grass, sage plenty. In two miles you arrive at river again	5—681
Creek, three feet wide. Road runs close to river..................	4—685
Bitter Cottonwood Creek. Some timber on it. After this the road leaves the river for six miles...............................	3—688
Road arrives at the river.......................................	6—694
Leave the old road and ford the river	1—695
Road turns between the two ridges, (ford twice)....................	1—696
Ford No. 4. Good camping place...................................	8—704
Pure water for sixteen miles.	
Ice Springs...	6—710
This is a low, swampy spot of land, on the right of the road. Ice may be found by digging down two feet. There are two alkali lakes a little further.	
Steep descent from the bluffs....................................	10—720
Ford of Sweet Water, No. 5. Grass and willow brush.............	1—721
Road joins the river and fords it.................................	4—725
Creek, two feet wide. Cold spring to the left......................	4—729
Road leaves the river. Good camping place........................	2—731
After this the road winds around and over a succession of hills and bottoms for three miles.	
Soft swamp and small. No place to camp	6—737
Strawberry Creek, five feet wide. Grass and willows..............	4—741
Branch of Sweet Water, two rods wide, two feet deep. Good place to camp. Wood and grass plenty................................	4—745
Willow Creek, eight feet wide, two feet deep. Good camping place. Ford three feet deep...	2—747
Sweet Water, three rods wide, three feet deep....................	5—752
South Pass, or summit of dividing ridge............................	9—761
Here the party leave the California trail, if they wish to go *via* Lander's Cut-off, and take the right hand road. They will find a small creek four miles from the post	4—765

Sweet Water, eight or ten miles.................................. 10—775
Small creek. Good feed and wood................................ 9—784
Big Sandy. Party will cross Little Sandy, and can camp on it....... 10—794
Leaving of Big Sandy. Party will cross rugged bluffs.............. 5—799

This stream, as well as all others here, abounds in trout. The writer caught one here that weighed two pounds.

Small creek. Good grass and water.............................. 8—807

The party is now between two ranges of snow-clad mountains, and may see Fremont's Peak by looking to the north.

Green River. Wood and grass. Here you ford the river........... 15—822
Second fork of Green River. Wood and grass plenty.............. 8—830
Leaving of Second fork. Here you will find a bad place to ford; do not go too low down to cross.................................. 5—835
Fourth fork of Green river...................................... 15—850

Here, a little west of the river, may be seen the grave of Martin Moran, who was killed by the Indians, the 18th of July, 1862.

Creek at base of mountains...................................... 9—859
Here you follow up a ravine, and camp on the creek, at the base of snow-clad hills... 6—865
Creek. Good feed, wood and water. (Steep canon to ascend)...... 11—876
Five miles good wood, water and grass........................... 5—881

The party is now on the Bear River chain of mountains.

Canon Creek. Heavy hills.. 14—895
Small creek, near Salt river. Heavy hills....................... 10—905

Here nearly all of our party got sick, and you will have to be careful of yourselves, as it rains every day and freezes at night.

Salt River.. 10—915
Mouth of canon.. 13—928
Creek at mouth of narrow canon. Here you will pass salt springs, where you may get as fine salt as can be found anywhere.......... 7—935
Beautiful valley. Water, wood and grass......................... 8—943
Cold spring in sight of river................................... 10—953
Small creek. Party will pass a large marsh. Here we run a huge grizzly into the cane-brakes and lost him........................ 10—963
Blackfoot Creek. Beautiful valley............................... 6—969

Party will have to go to the hill-sides to get fuel. Here the emigrant will take the right hand road for Snake River ferry.

Wolf Creek. Road rough until you reach Snake River bottom...... 8— 977
Luce's Creek. Wood, water and grass at all these creeks.......... 12— 989
Divide Creek.. 14—1003
John Gray's Creek... 12—1015
Snake River ferry, owned by Harry Rickard, who is very obliging, always ready to accommodate. This is the best ferry on the river.. 8—1023
When you leave Snake river, fill your kegs...................... 10—1033
Cumas Creek. Barren country,—no grass until you arrive at creek.. 24—1057
Scott's Ranche. Good feed, wood and water....................... 26—1083

You will now re-cross the Rocky mountains to the eastern side.

Summit Creek.. 9—1092

The party will take the right hand road between this camp and the next one, if you are going to Virginia City.

Johnston's Springs.. 20—1112
Camp on Read Rock... 14—1126
Horse Creek Ranche.. 23—1149
Bannock City.. 11—1160

ROUTE FROM DEER CREEK TO GALLATIN CITY.

Upon our map we have a route leading from the mouth of Deer Creek, which is some eighty miles west of Fort Laramie upon the Platte, to Gallatin City at the three forks of the Missouri. Mr. Bozeman, of Virginia City, surveyed the route last season, and will escort a train over the road the coming summer. To him we are indebted for the following table of distances:

From Deer Creek to Powder River	100	Miles.
" Powder River to Tongue River	85	"
" Tongue River to Rose Bud	20	"
" Rose Bud to Little Horn	15	"
" Little Horn to Big Horn	20	"
" Big Horn to Nez Perces Fork	40	"
" Nez Perces Fork to Clark's Fork	10	"
" Clark's Fork to Yellow Stone	15	"
Travel up the Yellow Stone	60	"
From leaving of the Yellow Stone to Gallatin River	16	"
Then down the Gallatin River to Gallatin City	40	"
From Deer Creek to Gallatin City	371	"
From Gallatin City to Virginia City	50	"
" Virginia City to Bannock City	70	"
" Gallatin City to Fort Benton	130	"

FROM SOUTH PASS TO SODA SPRINGS BY FORT BRIDGER.

TABLE OF DISTANCES.

South Pass	761	Black Foot, fourth time	2— 865
Pacific Springs	3—764	Small stream—swift current,	3— 868
Dry Sandy	11—775	Stream—good camping	4— 872
Junction of California and Oregon roads	6—781	Fort Bridger	8— 880
Little Sandy	8—789	Big Muddy	8— 888
Big Sandy	8—797	Little Muddy	12— 900
Big Sandy again	17—81	Good camping	30— 930
Green River ford	10—824	Cold Springs	15— 945
Road leaves Green river	5—829	Smith's Fork	8— 953
Black Foot Creek	15—844	Spring Fork	3— 956
Houris fork	4—848	Large stream	8— 964
Black Foot again	2—850	Willow Spring	24— 988
Small Creek	11—861	Halleck Fork	10— 998
Black Foot, third time	2—863	Quaking Asp	12—1010
		Soda Springs	15—1025

Here you strike the Salt Lake and Bannock route.

WESTHEIMER & EPPSTEIN,

(Successors to J. KINNEY & CO.,)

WHOLESALE DEALERS IN STAPLE & FANCY

GROCERIES,

LIQUORS,

SEGARS, TOBACCO, &c.,

No. 91 Felix Street,

ST. JOSEPH, MISSOURI.

Emigrants are requested to examine our stock before purchasing elsewhere.

C. D. SMITH & CO.,

WHOLESALE GROCERS

Importers, Distillers, Rectifiers,

AND DEALERS IN

FOREIGN AND DOMESTIC LIQUORS,

CIGARS, TOBACCO, PLAYING CARDS,

Oysters, Sardines, &c.,

89 Felix Street, between 3d and 4th,

ST. JOSEPH, MO.

HANNIBAL & ST. JOSEPH R. R.

AND CONNECTIONS.

THE ONLY ALL RAIL ROUTE

To the Upper Missouri River and the

FAR WEST!

The best Route from all EASTERN CITIES to the following Places, viz:

ALL RAIL TO } ST. JOSEPH AND ATCHISON
Iatan & Weston

7 Miles River Navigation to LEAVENWORTH,
40 " " " " KANSAS CITY and WYANDOTTE, by DAILY PACKET.

TO POINTS ABOVE ST. JOSEPH:

A splendid Packet leaves St. Joseph for Brownsville, Nebraska City, Plattsmouth, Council Bluffs, Omaha, &c.

DAILY OVERLAND MAIL STAGES

Leave terminus of Road for DENVER, SALT LAKE, WIND RIVER MINES, IDAHO and CALIFORNIA.

Daily Stages from LEAVENWORTH TO LAWRENCE, TOPEKA, and FT. RILEY.

Daily Stages from KANSAS CITY TO FT. SCOTT and SANTA FE.

☞CONNECTIONS FROM THE EAST, VIA CHICAGO and TOLEDO; Chicago, Burlington & Quincy, and Wabash Valley Railroads, make certain connections daily with

HANNIBAL & ST. JOSEPH RAILROAD

VIA ST. LOUIS.

TERRE HAUTE, ALTON & ST. LOUIS, OHIO & MISSISSIPPI, and CHICAGO, ALTON & ST. LOUIS RAILROADS make certain Connections with HANNIBAL & ST. JOE RAILROAD PACKET, leaving ST. LOUIS EVERY EVENING, making sure connections at HANNIBAL.

PASSENGERS TAKING NORTH MISSOURI R. R.

CONNECT CLOSELY AT MACON WITH H. & ST. JOE R.R. FOR THE WEST.

HANNIBAL & ST. JOSEPH R.R.

FREIGHT ARRANGEMENTS

Are perfected so fully that the Shipper can rely on QUICK TIME AND LOW RATES, with the assurance that all efforts will be used to deliver the Goods WITHOUT OVER-CHARGE, and in as good condition as received. Through Freight contracts made at Offices of all connecting Roads in the East, and at the Offices of the Company.

THROUGH TICKETS FOR SALE, VIA HANNIBAL & ST. JOSEPH R. R., at all PRINCIPAL TICKET OFFICES IN THE EAST.

J. T. K. HAYWARD, Gen'l Sup't, Hannibal, Mo.

W. P. ROBINSON, Gen'l Freight Agent. P. B. GROAT, Gen'l Ticket Ag't, Hannibal, Mo.

J. Q. A. BEAN, Eastern Agent, Boston. HENRY STARRING, Gen'l Agent, Chicago.

John L. Campbell

WHO IS GOING TO THE

GOLD MINES

OF

Colorado and Idaho?

The Best Way to get there from the States.

That route that can take you the farthest West by Railroad must certainly be the best, because it will lessen the time of going and the fatigue of the journey.

The Route that can do this is the

Chicago, Burlington & Quincy R. R.

In connection with the

Hannibal & St. Joseph R. R.

This line extends from

CHICAGO TO ST. JOSEPH & ATCHISON,

The Starting Point of the Great Over-Land Stage Company's Six-Horse Coaches for

DENVER, IDAHO, SALT LAKE & CALIFORNIA,

Only 24 to 30 hours' Time

is required to go from Chicago to St. Joseph or Atchison on the Missouri river.

RATES OF FARE.

Chicago to St. Joseph..........................	Seventeen Dollars.
Chicago to Atchison.............................	Eighteen Dollars.

During the season of navigation on the Missouri river, steamboats run regularly from St. Joseph to Omaha City and Council Bluffs, touching at intermediate points, thus making a continuous route from Chicago to any point on the Upper Missouri River.

Two Passenger Trains

leave Chicago daily on the arrival of trains from the East.

No other railroad than the Hannibal and St. Joseph extends through to the Missouri River, and none other than the Chicago, Burlington and Quincy runs in direct connection with the Hannibal and St. Joseph, from Chicago.

Take the route that extends farthest West, and save Time and Money.

Through Tickets to the Missouri river for sale at all the principal ticket offices in the East, and at the office of the Chicago, Burlington and Quincy R. R., in the Great Central Depot, foot of Lake Street, Chicago.

Buy Through Tickets, by all means.

SAMUEL POWELL,
Gen'l Ticket Agent.

C. G. HAMMOND,
Gen'l Sup't

WM. H. YOUNG,

GENERAL

FEED & SALE STABLE,

Fifth Street, opposite Gas Works,

St. Joseph, Mo.

Persons fitting out for the New Mines will find it to their interest to give me a call, as my stable is situated in a central part of the city, convenient to business and the Hotels.

MULES AND HORSES

Constantly on hand, and for sale at the lowest market price.

REFERENCES.

Gurney & Co.,	Wholesale Boots and Shoes,	St. Joseph.
Jno. Calhoun,	Banker.	"
Woolworth & Barton,	General Freighters,	"
Jno. L. Bittinger,	Postmaster,	"
Hawkins, Smith & Co.,	General Commission,	Chicago.
Jno. L. Noble,	" "	"
R. R. & C. W. Day,	Stoves and Tinware,	"

PATEE HOUSE,

St. Joseph, Missouri.

This Hotel, capable of accommodating, comfortably, 350 guests, is surpassed by none in the West, is within four minutes' ride of the Depot of the Hannibal & St. Joseph and Platte County Railroads, where Omnibusses are always in attendance to convey passengers to and from the Hotel.

The Table is at all times supplied with the Choicest Luxuries from Home and Foreign Markets.

Connected with the Hotel is a

Steam Laundry,

Also, WARM AND COLD BATHS.

The House and Fixtures are entirely new, and no pains will be spared by the Proprietor to constitute it in every respect a FIRST CLASS HOTEL.

T. M. ESPY, Proprietor.

JOHN R. WALSH,

WHOLESALE

NEWS AGENT

AND

BOOKSELLER,

MADISON STREET,

Corner of Custom House Place,

CHICAGO, - - - - - - ILLINOIS.

WESTERN DEALERS

SUPPLIED WITH EVERYTHING IN HIS LINE AT THE EARLIEST MOMENT.

PRICE LISTS FREE, ON APPLICATION.

PERIODICAL DEALERS

In Illinois, Wisconsin, and Iowa, will find it to their interest to send for one of his Price Lists.

John L. Campbell

Index

A

Alkali Lake 58
Alleghany Mountains 8
Allan, J.F. 40, 41
Allen, Mr. 26
America 8
American Desert 7
American Fur Company 17,21
Ancient mines 29
Antelope 58
Ash Hollow 61
Ash Point 55
Atchinson, Kansas 53
Atlantic slope 11,13
Auxiliary services 47

B

Baerman, Fred 67
Balder Creek 31
Bannock City (Idaho City) 7, 12, 13, 16, 21, 23, 26, 27, 28, 29, 30, 31, 32, 37, 38, 41, 59, 64, 65, 74
Bannock Road 11, 59
Bannock Indians 35
Barranca Mountains 8
Barton, Mr. 31
Baugh, Mr. 29
Bear River 11, 12, 59
Bear River ferry 59
Bear River Mountains 8, 9
Beaver Creek 58
Beranger 52
Bernard & Redman 73
Beven's Gulch 31
Big Bend 58
Big Hole River 32
Big Horn Mountains 24, 32, 33
Big Horn River 32
Bijou 58
Bitter-Root Mountains 19
Bitter-Root River & Valley 18, 19, 20, 21, 41
Black Buttes 59
Black, Capt. 12
Black Hills 7, 24
Blackfoot Creek 59, 64, 65
Blackfoot Indians 35
Blackfoot River 11, 18, 26
Blue River, Little 50
Blue River, Big 52, 55
Boise River 26
Boner 58
Boon's 58
Box Elder 59
Brewery 59
Bridger's Pass 59
Brigham City 59
British America 5,41
Brown's Bridge 52
Brown, Geo. W. 73
Brown's Gulch 31
Buffalo chips 9

C

California 6, 11, 15, 25, 26, 35, 37, 49, 63
California, Gulf of 41
California troops 11
California Road 65
Canon Creek 59
Carter, Mr. 28
Cascade Range 20
Cascades, The 26
Central City, C.T. 66
Cherokee 58
Chicago, Ill. 6, 74, 79
Chicago, Burlington & Quincy RR. 76
Child's 58
Chimney Rock 62
Chrisman, Geo. 28
Christianity 19
Church Buttes 59
Clark's Fork River 41, 65
Clearwater River (Kooskooskia) 5, 25
Clothing 45
Coeur d' Alene Mission 19

Cold Springs 58
Colorado 5, 6, 32, 49
Colorado River 41
Columbia River 7, 19, 26, 41
Columbus, Neb. 57
Congress, U.S. 55
Congress, Organic Act 1863 5
Connor, Gen. 12
Constable, D.A. 71
Cooper Creek 59
Cotton-wood City, Mont. 20
Cottonwood Creek 59
Cottonwood Springs 58
Cottonwood Ranch 55
Council Bluffs, Iowa 51, 78
Craig 58

D

Dakota lode 26, 29
Dakota Territory 5, 6
Dalles, The 26
Daniel's 59
Davenport, Chas. W. 71
Deer Lodge, Mont. 18, 20
Deer Lodge River 18, 21, 32
Delaware Indians 55
Dennison's 58
Denver, Colo. 53, 58
Detroit, Mich. 74
Devil's Gate 63
Diamond Springs 58
Dodge County, Neb. 57
Dow, Mr. 12
Drift mining 37, 39
Dry diggings 38
Duck Lake 59
Dug Springs 59
Dust prices 31

E

Eagle's Nest 58
Echo Canyon 59
Eckhart, Wm. 72
Edgerton, Chief Justice 27
El Dorado 35
Elk City, Ida. 25
Elk City, Neb. 57
Elk Mountain 59
Elkhorn River 57, 58
Elm Creek 55
Elwood, Neb. 49
Emigrant's Guide 60-65

F

Fairweather Gulch 30, 31
Fairweather, Mr. 30
Farmington 59
Fisk, Capt. 13, 15
Flat-head Indians 19, 20, 35
Flathead River 26
Florence City, Ida. 25
Flumes 26
Fort Benton 17, 23, 26, 65
Fort Bridger 11, 59, 65
Fort Connor 12
Fort Hall 19
Fort Kearney 7, 9, 49-50, 51, 52, 55, 57, 58
Fort Laramie 7, 9, 17, 18, 41, 62
Fort Leavenworth 17, 18
Fort Lupton 58
Fort Snelling 17, 18
Fort Union 21
Fourth of July 11
Franklin 59
Fremont, Capt. John C. 55
Fremont, Neb. 57
Fremont Orchard 58
Fremont Springs 58
Friedenheit, G. 70

G

Galena, Ill. 29
Gallatin City, Mont. 23, 31, 65
Gallatin River 21, 65
Gem of the Mountains 5
Gilman's 58
Gold mines 25-33
Gold price 25
Gold yield 25-28
Grand Island, Neb. 57, 61
Grant, John 21
Grasshopper Creek 26, 29
Grasshopper mines 26, 37
Grayson's 50
Great Desert 30
Great Lakes 18
Gree-backs 31

Green River . . . 59, 64, 65
Grizzly bear . . . 9, 10, 11, 45
Guittard's . . . 50
Gulch mining . . . 41
Gurney & Co. . . . 66

H

Hackley, Samuel . . . 27
Hallet & Co. . . . 55
Ham's Fork . . . 59
Hanging Rock . . . 59
Hannibal & St. Joseph RR. . . . 49, 51, 74, 75
Harris' Gulch . . . 31
Hell Gate City, I. T. . . . 18
Hell Gate River . . . 18, 19, 21, 41
Holliday's Stage Line . . . 53, 58
Hooks . . . 50
Horn River, Big . . . 65
Horn River, Little . . . 65
Horse Creek mines . . . 29
Hudson Bay . . . 41
Hudson River . . . 8
Hunkins, Col. . . . 29
Huron . . . 55
Hydraulic mining . . . 36

I

Idaho . . . 5, 6, 9, 11, 12, 13, 15, 17-24, 25, 32, 41, 43, 47, 49, 51, 53
Idaho City (formerly Bannock) . . . 12
Illinois . . . 57
Independence Rock . . . 63
Independence Valley . . . 11, 59
Indians . . . 6, 19, 20, 32, 35, 43, 45
Iowa . . . 55
Iowa Cavalry, 7th. . . . 57

J

Jeff Davis' Gulch . . . 29
Jefferson River . . . 21, 26, 30, 31
Jesuit priests . . . 19-20
Julesburg . . . 58

K

Kahh & Co. . . . 70
Kansas River . . . 55
Kearney City, Neb. . . . 57
Keith & Snell . . . 72
Kickapoo . . . 50
Kelly's . . . 58
Kimball, Utah . . . 59
Kinnekuk . . . 50, 55
Kiowa . . . 50
Knowles, John . . . 27
Koch & Borngesser . . . 69
Kooskooskia River . . . 5, 41

L

Laclede . . . 59
Lancaster . . . 55
Lander, Col. . . . 11
Lander's Cut-off . . . 8, 9, 63
Laporte . . . 58
Laramie Creek . . . 50, 59
Laramie Peak . . . 62
Laramie River . . . 7, 59
Latham . . . 58
Lawrence, Neb. . . . 55
Leavenworth, Kansas . . . 53, 55
Leavenworth & Pawnee RR. Co. . . . 55
Locklands . . . 55
Les Montagnes Rocheuses . . . 8
Lewiston, Ida. . . . 25
Liberty Farm . . . 50
Lone Tree Ranch . . . 50, 59, 60
Loup Fork . . . 57, 60
Lovejoy, Mr. . . . 28
Log Chain . . . 50

M

McCord, Nave & Co. . . . 78
McDonnel, Mr. . . . 28
McGibbons, Sam S. . . . 67
McGregor, Iowa . . . 26
Madison River . . . 21, 30
Malheur River . . . 35
Marysville . . . 55
Maxwell & Warfield . . . 72
Medicine Bow . . . 59
Mexico . . . 29
Mexican Chronicles . . . 29-30
Mexico, Gulf of. . . . 41
Michigan Central & Great Western RR. . . . 74
Michigan Wagon Depot . . . 80
Midway . . . 58
Millersville . . . 59

Minnehaha 55
Minnesota 30
Mississippi River 7, 8, 17, 19
Missouri (State) 55, 57
Missouri Pacific RR. 55
Missouri River 13, 17, 20, 21, 22, 23, 25, 26, 29, 32, 41, 45, 49, 51, 52, 53, 55, 57
Mormons 11, 23
Morrisite Faction 11, 12
Mountain Dell 59
Muddy River 59, 65
Mullan, Lieut. John 19-20
Murray, Mr. 28

N

Nebraska City, N. T. 52
Nebraska Territory 5, 55
Neddle Rock 59
Nemaha River, Little 52
Niagara 8
Nevada City 30
New England 6
New York State 6, 55
Nez Perce Indians 35
North America 41
North Side Ditch Co. 39
North-West, The 18
North-West Ditch Co. 78
North-Western Fur Co. 78

O

Ogden, Utah 59
Ohio River 8
Ohio Troops 7
Okaw 50
Olatha 52
Oliver & Co. 28
Olympia, Washington 19
Omaha, Neb. 29, 51, 52, 57, 60, 78
Onondaga, N. Y. 9
Ontario, Lake 8
Ophir 35
Oregon 5, 6, 20, 25, 35
Oregon Trail 8, 11
Oro Fino, Ida. 25
Otoe 50
Outfits . . . . 43; costs 43, 45
Outfitting points 49-58
Overland Telegraph 58
Owen, Jonah 20

P

Pacific Coast 6, 13, 17, 20, 25, 35
Pacific Ocean 7, 41
Pass Creek 59
Patee House 77
Pawnee 50
Pay-dirt 27-28
Peabody & Co. 31
Penick & Loving 69
Pierson's 58
Pine Grove 59
Platte River 7, 11, 32, 51, 55, 57, 58, 59, 61, 62
Platte Station 58
Plum Creek 58
Pony 45
Potawatomie Indians 55
Powder River 65
Prairie Lake 52
Prickly Pear Creek 32
Provisions, cost of 28, 31, 43, 45

Q

Quaking Asp Springs 59, 65
Quartz mining 41-42

R

Raferty & Kelly Co. 67
Richardson, W. K. 71
Rickard's Ferry 13, 14, 64
Rivers 41-42
Rock Creek 55, 59
Rock Point 59
Rock Springs 59
Rocky Mountains 5, 7, 13, 19, 20, 24, 41

S

St. Joseph, Mo. 49-50, 66, 67, 68, 69, 70, 71, 72, 73, 77, 78, 80
St. Louis, Mo. 7, 23, 49
St. Mary's River 18
St. Mary's Village 19
Sage Creek 59
Salmon River 25, 32, 41
Salt Lake 11, 12
Salt Lake City 26, 28, 53, 58, 59

Salt Wells 59
Sand Hill Pond 55, 58
Sandy River, Big 50, 55, 65
Sandy River, Little 55, 65
Saskatchewan River 32, 41
Saunders House 67
Scotchman 12
Scott's Bluffs 62
Scurvy 46
Seneca 50
Sicefuff, W. D. 72
Sluices 34, 38
Sluicing 37-38
Smith, C. D. 68
Smith, Joe 12
Snake Indinas 35
Snake River 5, 13, 14, 25, 32, 41, 64
Soda Springs 11, 12, 59, 65
South Pass 7, 8, 9, 11, 13, 65
Spaniards 29
Spring Hill 58
Sulphur Springs 59
Stage coaches 28, 30-31
Stamping mill 40-42
Stevens Cantonment 18
Stevens, Gov. I. I. 17, 19-20
Stinking Water 28, 30, 31
Stove, emigrant 44
Summit 50
Summit City 30
Summit Creek 59
Superior, Lake 8
Sweet Water River 63, 64

T

Table Mountains 8
Telegraph 28, 58
Tents 44
Thirty-two Mile Creek 50, 55
Thompson's 50
Thompson River, Little 58
Thompson River, Big 58
Three Forks Valley 23
Tongue River 65
Tootle & Fairleigh 73
Turner, Frazer & Williams 73

U

Union Pacific RR. 51, 55, 57
United States Government 15, 17, 28
United States Troops 12
Upas 12
Utah 6, 11, 12, 44

V

Valley Station 58
Vanderbert, Mr. 32
Vermillion River, Black 55
Virginia City, I. T. 21, 26, 30, 65
Virginia Dale 58

W

Wages 28, 31
Walnut Creek 55
Washington Territory 5, 6, 25
Waskie 59
Weber, Utah 59
Weil, L. 70
Wellsville 59
West, The 8
West Ditch Co. 38
Western Continent 8
Western Stage Co. 57
Westheimer & Epptein 68
Willow Island 58
Willow Springs 59
Wilson & Co. 21
Wilson's Bridge 52
Wind River Mountains 8, 24
Winnipeg, Lake 32, 41
Woolworth & Barton 69
Woolworth & Co. 71
Wyeth, W. M. 70

Y

Yellow Stone Mountains 24
Yellow Stone River 21, 23, 25, 32, 33, 41, 65
Yields, Gold 27
Young, Brigham 12, 23
Young, Wm. H. 77

COLOPHON

The John Lyle Campbell book, Six Months in the New Gold-Diggings, *was printed in the workshop of Glen Adams, which is located in the quiet country village of Fairfield, Washington, southern Spokane County, and on the north edge of the Palouse wheat farming district. The book was set by Miss Bobi Pearson using a Compugraphic 48 photosetter. The setting was in twelve point Baskerville which was given a fifteen per cent enlargement with the camera before printing. The running heads were hand set using eighteen point Bologna hand type from the Stephenson, Blake foundry in Sheffield, England. Page numbers are Baskerville. The sheets were printed by Ralph Decker using a Hamada 770CD offset press. The printing of the title page, colored initials, vignettes and general book design was by Glen Adams. Camera-darkroom work was by Evelyn Clausen. The book was stripped by Mike Brunk. Indexing was by Edward J. Kowrach. The sheets were folded and signatures assembled by Evelyn Clausen, Carol Binder and Edward Kowrach. Paper stock is eighty pound Lowell, a Zellerbach sheet. This was a fun project. We had no special difficulty with the work.*